WONDERS OF THE UNIVERSE

For Mum, Dad, and Sandra—none of this would have been possible without you.
Brian Cox

For my dad, Geof Cohen (1943–2007)
Andrew Cohen

First published in the United Kingdom by Collins, an imprint of HarperCollins*Publishers*, by arrangement with the BBC.

HarperCollins books may be purchased for educational, business, or sales promotional use. For information please write: Special Markets Department, HarperCollins*Publishers*, 10 East 53rd Street, New York, NY 10022.

FIRST EDITION

Designed by Studio8 Design

Printed on acid-free paper

Library of Congress Cataloging-in-Publication Data has been applied for.

ISBN 978-0-06-211054-1

11 12 13 14 15 SCP 10 9 8 7 6 5 4 3 2 1

WONDERS OF THE UNIVERSE

BRIAN COX
AND ANDREW COHEN

HARPER DESIGN
An Imprint of HarperCollinsPublishers

CHAPTER 1

MESSENGERS

CHAPTER 2

STARDUST

CHAPTER 3

FALLING

CHAPTER 4

DESTINY

G-STAR

INTRODUCTION

THE UNIVERSE

At 13.7 billion years old, 93 billion light years across and filled with 100 billion galaxies – each containing hundreds of billions of stars – the Universe as revealed by modern science is humbling in scale and dazzling in beauty. But, paradoxically, as our knowledge of the Universe has expanded, so the division between us and the cosmos has melted away. The Universe may turn out to be infinite in extent and full of alien worlds beyond imagination, but current scientific thinking suggests that we need it all in order to exist. Without the stars, there would be no ingredients to build us; without the Universe's great age, there would be no time for the stars to perform their alchemy. The Universe cannot be old without being vast; there may be no waste or redundancy in this potentially infinite arena if there are to be observers present to gaze upon its wonders.

The story of the Universe is therefore our story; tracing our origins back beyond the dawn of man, beyond the origin of life on Earth, and even beyond the formation of Earth itself; back to events – perhaps inevitable, perhaps chance ones – that occurred less than a billionth of a second after the Universe began.

AN ANCIENT WONDER

On Christmas Eve 1968, Apollo 8 passed into the darkness behind the Moon, and Frank Borman, Jim Lovell and William Anders became the first humans in history to lose sight of Earth. When they emerged from the Lunar shadow, they saw a crescent Earth rising against the blackness of space and chose to broadcast a creation story to the people of their home planet. A quarter of a million miles from home, lunar module pilot William Anders began:

'We are now approaching lunar sunrise and, for all the people back on Earth, the crew of Apollo 8 has a message that we would like to send to you.
In the beginning God created the heaven and the Earth.
And the Earth was without form, and void; and darkness was upon the face of the deep.
And the Spirit of God moved upon the face of the waters. And God said, Let there be light: and there was light.
And God saw the light, that it was good: and God divided the light from the darkness.'

The emergence of light from darkness is central to the creation mythologies of many cultures. The Universe begins as a void; the Maori called it *Te Kore*, the Greeks *Chaos*. The Egyptians saw the time before creation as an infinite, fathomless ocean out of which the land and the gods emerged. In some cultures, God is eternal: He created the Universe out of nothing and will outlast it. In others, such as some Hindu traditions, a vast primordial ocean predates the heavens and Earth. Lord Vishnu floated, asleep, on the ocean, entwined in the coils of a giant cobra, and only when light appeared and the darkness was banished did he awake and command the creation of the world.

We still don't know how the Universe began, but we do have very strong evidence that something interesting happened 13.75 billion years ago that can be interpreted as the beginning of our universe. We call it the Big Bang. (We must

The cosmos is about the smallest hole that a man can hide his head in.

— G.K. Chesterton

be careful with our choice of words here, because this is a book about science, and the key to good science is the separation of the known from the unknown.) This interesting thing that happened corresponds to the origin of everything we can now see in the skies. All the ingredients required to build the hundreds of billions of galaxies and thousands of trillions of suns were once contained in a volume far smaller than a single atom. Unimaginably dense and hot beyond comprehension, this tiny seed has been expanding and cooling for the last 13.75 billion years, which has been sufficient time for the laws of nature to assemble all the complexity and beauty we observe in the night skies. These natural processes have also given rise to Earth, life, and also consciousness, which in many ways is harder to comprehend than the mere emergence of the seemingly infinite stars.

Care is in order, because the very beginning – by which we mean the events that happened during the Planck epoch – the time period before a million million million million million million millionths of a second after the Big Bang, is currently beyond our understanding. This is because we lack a theory of space and time before this point, and consequently have very little to say about it. Such a theory, known as quantum gravity, is the holy grail of modern theoretical physics and is being energetically searched for by hundreds of scientists across the world. (Albert Einstein spent the last decades of his life searching in vain for it.) Conventional

When you set sail for Ithaca, wish for the road to be long, full of adventures, full of knowledge.

– C. P. Cavafy

thinking holds that both time and space began at time zero, the beginning of the Planck era. The Big Bang can therefore be regarded as the beginning of time itself, and as such it was the beginning of the Universe.

There are alternatives, however. In one theory, what we see as the Big Bang and the beginning of the Universe was caused by the collision of two pieces of space and time, known as 'branes', that had been floating forever in an infinite, pre-existing space. What we have labelled the beginning was therefore nothing more significant than a cosmic collision of two sheets of space and time.

It may be that the question 'Why is there a Universe?' will remain forever beyond us; it may also be that we will have an answer within our lifetimes, but the quest has to date proved more valuable than the answer because the ancient search for origins lies at the very heart of science. Indeed, it lies at the heart of much of human cultural development. The desire to understand events beyond the terrestrial seems to be innate, because all the great civilisations of antiquity have shared it, developing stories of beginnings, origins and endings. It is only recently that we have discovered that this quest is also profoundly useful in a practical sense. When coupled with the scientific method, this quest has allowed us not only to better understand nature, but to manipulate and control it for the enrichment of our lives through technology. The well-spring of all that we take for granted, from medical science to intercontinental air travel, is our curiosity.

THE VALUE OF WONDER

The idea that a journey to the edge of the Universe is deeply relevant to our everyday lives lies at the heart of *Wonders of the Universe*. I cannot emphasize enough my strong conviction that exploration, both intellectual and physical, is the foundation of civilisation. So whilst building rockets to the Moon and telescopes to capture the light from the most distant stars may seem like an interesting luxury, such a view would be superficial, incorrect and downright daft – to borrow a phrase from my native Oldham. We are part of the Universe; its fate is our fate; we live in it and it lives in us. How can anything be more important, relevant and useful than understanding its workings?

When we began to think about the series, we wanted to make programmes that were more than a simple tour of the wonders of the Universe. Of course black holes, colliding galaxies and stars at the edge of time are fascinating, and we see them all, but to characterise the ancient science of

PREVIOUS SPREAD AND LEFT: The work of the space programmes across the world cannot be viewed as a luxury, but rather as a necessity. It is through the missions of space shuttles such as Atlantis (left) and Endeavour (previous page) that we can truly begin to understand the origins and the workings of the Universe and use this information to plan for our future on Earth.

astronomy as a spectator sport would be to miss the point. The wonders we see through our telescopes are laboratories where we can test our understanding of the natural world in conditions so extreme that we will never be able to recreate them here on Earth. With this in mind, we decided to base the programmes around scientific themes rather than the wonders themselves.

'Messengers' is about light – our only connection with the distant Universe that may lie forever beyond our reach. But it is also about the information stored within light itself, and how that information got there; the fingerprints of the chemical elements are to be found in the most distant starlight, enabling us to know with certainty the composition of the most distant star.

'Stardust' asks an ancient question: what are the building blocks of the Universe? And also, how was the raw material of a human being assembled from the debris of the Big Bang? – a searingly hot, yet beautifully ordered fireball with no discernable structure.

'Falling' tells the story of the great sculptor of the Universe: gravity. For some reason that we do not understand, gravity is by far the weakest of the four fundamental forces in the Universe, but because it has an infinite range and acts between everything that exists, its influence is all-pervasive. Our most precise theory of gravity, Einstein's General Theory of Relativity, dates from 1915, which makes it the oldest of the modern theories of the forces. The theory of the electromagnetic force, Quantum Electrodynamics, dates from the 1950s, while the theory of the Strong Nuclear Force from the 1960s and 70s. Our description of last of the four, the Weak Nuclear Force, resides in the Standard Model of particle physics. This theory, a product of the 1970s, unifies the description of the Weak Nuclear Force with Quantum Electrodynamics, although there is a missing piece of the theory known as the Higgs Boson that is currently being searched for at the Large Hadron Collider at CERN in Geneva. Until the Higgs Boson, or whatever does its job, is found, we cannot claim to have a working description of the Weak Nuclear Force and its relationship with electromagnetism.

However despite the long pedigree and beautiful accuracy and elegance of Einstein's theory of gravity, it is known to be incomplete. Our description of the Universe breaks down in the heart of its most evocatively named wonders. Black holes are known to exist at the centre of galaxies such as the Milky Way, and are dotted throughout the cosmos; the carcasses of the most massive stars in the

RIGHT: Armed with a greater knowledge and understanding of our universe, and also with new technology and modern approaches to science, we can discover wonders of the Universe that would have remained hidden to us centuries ago. Galaxies such as the spiral-shaped Dwingeloo 1 have recently been found hidden behind the Milky Way. This discovery supports what we already know: that there are many more wonders out there in the Universe that we have yet to discover.

Universe. We see them by their influence on passing stars and by detecting the intense radiation emitted by gas and dust that has the misfortune to venture too close to their event horizons. We have even seen their formation in the most violent cosmic events – supernova explosions. These events mark the destruction of stars that once burned brightly for millennia, completed in a matter of minutes.

The final chapter, 'Destiny', delves into the distant past and the far future; following the inevitable ticking of the universal clock. It is also the chapter that most directly touches on the great contribution of engineering to our story. The science of thermodynamics, which has become our guide to the ultimate fate of the Universe, arose from considerations of the efficiency of steam engines in the nineteenth century and not a desire to peer out towards a possibly infinite future. In 'Destiny' we describe thermodynamics in detail, and show how this quintessentially nineteenth-century science allows us to speculate with some grounding in reality about events that will happen 10,000,000,000,000,000,000,000,000,000 ,000,000,000,000,000,000,000,000,000,000,000,000,000, 000,000,000,000,000,000,000,000,000,000,000 years from now. Not bad for the pioneers of the age of steam.

So as we look to the future and survey the wonders of our universe, we discover that Einstein's theory of gravity, our best description of the fabric of the Universe, predicts its demise inside black holes. The collapsing remnants of the most luminous stars represent the edge of our understanding of the laws of physics and therefore the edge of our understanding of the wonders of the Universe. This is exactly where every scientist wants to be. Science is a word that has many meanings; one might say science is the sum total of our knowledge of the Universe, the great library of the known, but the practice of science happens at the border between the known and the unknown. Standing on the shoulders of giants, we peer into the darkness with eyes opened not in fear but in wonder. The fervent hope of every scientist is that they glimpse something that not only requires a new scientific theory, but that requires the old theory to be replaced. Our great library is constantly being rewritten; there are no sacred tomes; there are no untouchable truths; there is no certainty; there is simply the best description we have of the Universe, based purely on our observations of its wonders.

The scientific project is ultimately modest: it doesn't seek universal truths and it doesn't seek absolutes, it simply seeks to understand – and therein lies its power and value. Science has given us the modern world, of that there can be no doubt. It has improved our lives beyond measure; increased life expectancy, decreased child mortality, eradicated many diseases and rendered many more impotent. It has given many of us the gift of time, freed us from the drudgery of mere survival and allowed us to open our minds and explore. Science is therefore a virtuous circle; its discoveries creating more time and wealth that we can, if we are wise, invest in further voyages of exploration and discovery. But for all its undoubted usefulness, I maintain that science is fuelled not by utilitarian desire but by curiosity. The exploration of the Universe and its wonders is as important as the search for new medical treatments, new energy sources or new technologies, because ultimately all these valuable advances rest on an understanding of the basic laws that govern everything in nature, from atoms to black holes and everything in between. This is why curiosity-driven science is the most valuable of pursuits, and this is why we must continue our journey into the darkness ◉

CHAPTER 1

MESSENGERS

THE STORY OF LIGHT

Throughout recorded history humans have looked up to the sky and searched for meaning in the heavens. The science of astronomy may now conjure thoughts of telescopes and planetary missions, but every modern moment of discovery has a heritage that stretches back thousands of years to the simplest of questions: what is out there? Light is the only connection we have with the Universe beyond our solar system, and the only connection our ancestors had with anything beyond Earth. Follow the light and we can journey from the confines of our planet to other worlds that orbit the Sun without ever dreaming of spacecraft. To look up is to look back in time, because the ancient beams of light are messengers from the Universe's distant past. Now, in the twentieth century, we have learnt to read the story contained in this ancient light, and it tells of the origin of the Universe.

Karnak Temple, home of Amun-Re, universal god, stands facing the Valley of the Kings across the Nile in the city of Luxor. In ancient times Luxor was known as Thebes and was the capital of Egypt during the opulent and powerful New Kingdom. At 3,500 years old, Karnak Temple is a wonder of engineering, with thousands of perfectly proportioned hieroglyphs, and an architectural masterpiece of ancient Egypt's golden age; it is a place of profound power and beauty. Ten European cathedrals would fit within its walls; the Hypostyle Hall alone, an overwhelming valley of towering pillars that once held aloft a giant roof, could comfortably contain Notre Dame Cathedral.

Religious and ceremonial architecture has had many functions throughout human history. There is undoubtedly a political aspect – these monumental edifices serve to cement the power of those who control them – but to think of the great achievements of human civilisation in these terms alone would be to miss an important point. Karnak Temple is a reaction to something far more magnificent and ancient. The scale of the architecture forcibly wrenches the mind away from human concerns and towards a place beyond the merely terrestrial. Places like this can only be built by people who have an appropriate reverence for the Universe. Karnak is both a chronicle in stone and a bridge to the answer to the eternal question: what is out there? It is an observatory, a library and an expression carved out of the desert of cosmological curiosity and the desire to explore.

Egyptian religious mythology is rich and complex. With almost 1,500 known deities, countless temples and tombs and a detailed surviving literature, the mythology of the great civilisation of the Nile is considered the most sophisticated religious system ever devised. There is no such thing as a single story or tradition, partly because the dynastic period of Egyptian civilisation waxed and waned for over 3,000 years. However, central to both life and mythology are the waters of the Nile, the great provider for this desert civilisation. The annual floods created a fertile strip along the river that is strikingly visible when flying into Luxor from Cairo, although since 1970 the Aswan Dam has halted the ancient cycle of rising and falling waters and today the verdant banks are maintained by modern irrigation techniques. The rains still fall on the mountains south of Egypt during the summer, and before the dam they caused the waters of the Nile to rise and flood low-lying land until they cease in September and the waters recede, leaving life-giving fertile soils behind.

The dominance of the great river in Egyptian life, unsurprisingly, found its way into the heart of their religious tradition. The sky was seen as a vast ocean across which the gods journeyed in boats. Egyptian creation stories speak of an infinite primordial ocean out of which a single mound

PREVIOUS PAGE: The spectacular remains and towering pillars of Karnak Temple are a testament to the Egyptian belief in the power and importance of the Amun-Re, the Sun God, in their daily life, and of the Sun itself.

LEFT: The power of the supreme god Amun-Re is felt everywhere at Karnak. Representations of him cover the walls; the carvings mostly depict him as human with a double-plumed crown of feathers alongside the Pharoah, but also in animal form as a ram.

BELOW AND RIGHT: The location and alignment of this impressive building, like everything else about it, has meaning. Egyptologists have evidence to support their belief that it was constructed as a sort of calendar; two columns frame the light of the sun as it rises on the winter solstice.

of earth arose. A lotus blossom emerged from this mound and gave birth to the Sun. In this tradition, each of the primordial elements is associated with a god. The original mound of earth is the god Tatenen, meaning 'risen land' (he also represented the fertile land that emerged from the Nile floods), while the lotus flower is the god Nefertem, the god of perfumes. Most important is the Sun God, born of the lotus blossom, who took on many forms but remained central to Egyptian religious thought for over 3,000 years. It was the Sun God who brought light to the cosmos, and with light came all of creation.

At Karnak, the Sun God reigns supreme as Amun-Re, a merger between the god Amun, the local deity of Thebes, and the ancient Sun God, Re. This tendency to merge gods is widespread in Egyptian mythology, and with the mergers comes increasing theological complexity. Amun can be seen as the hidden aspect of the Sun, sometimes associated with his voyage through the Underworld during the night. In the Egyptian *Book of the Dead,* Amun is referred to as the 'eldest of the gods of the eastern sky', symbolising his emergence as the solar deity at sunrise. As Amun-Re, he became the King of the Gods, and as Zeus-Ammon he survived into Greek and Roman times. Worship of Amun-Re as the supreme god became so widespread that the Egyptian religion became almost monotheistic during the New Kingdom. Amun-Re was said to exist in all things, and it was believed that he transcended the boundaries of space and time to be all-seeing and eternal. In this sense, he could be seen as a precursor to the gods of the Judeo-Christian and Islamic traditions.

The walls of Karnak Temple are literally covered with representations of Amun-Re, usually depicted in human form with a double-plumed crown of feathers – the precise meaning of which is unknown. He is most often seen with the Pharaoh, but he also appears at Karnak in animal form, as a ram.

The most spectacular tribute of all to Amun-Re, though, lies in Karnak's orientation to the wider Universe. The Great Hypostyle Hall, the dominant feature of the temple, is aligned such that on 21 December, the winter solstice and shortest day in the Northern Hemisphere, the disc of the Sun rises between the great pillars and floods the space with light, which comes from a position directly over a small building inside which Amun-Re himself was thought to reside. Standing beside the towering stone columns watching the solstice sunrise is a powerful experience. It connects you directly with the names of the great pharaohs of ancient Egypt, because Amenophis III, Tutankhamen and Rameses II would have stood there to greet the rising December sun over three millennia ago.

The Sun rises at a different place on the horizon each morning because the Earth's axis is tilted at 23.5 degrees to the plane of its orbit. This means that in winter in the Northern Hemisphere the Earth's North Pole is tilted away from the Sun and the Sun stays low in the sky. As Earth moves around the Sun, the North Pole gradually tilts towards the Sun and the Sun takes a higher daily arc across the sky until midsummer, when it reaches its highest point. This gradual tilting back and forth throughout the year means that the point at which the Sun rises on the eastern horizon also moves each day. If you stand facing east, the most southerly rising

point occurs at the winter solstice. The sunrise then gradually drifts northwards until it reaches its most northerly point at the summer solstice. The ancients wouldn't have known the reason for this, of course, but they would have observed that at the solstices the sunrise point stops along the horizon for a few days, then reverses its path and drifts in the other direction. The solstices would have been unique times of year and important for a civilisation that revered the Sun as a god.

Standing in Karnak Temple watching the sunrise on this special midwinter day the alignment is obvious, but proving that ancient sites are aligned with events in the sky is difficult and controversial. This is because a temple the size of Karnak will always be aligned with something in the sky, simply because it has buildings that point in all directions! However, a key piece of evidence that convinced most Egyptologists that Karnak's solstice alignment was intentional concerns the two columns on either side of the building in which Amun-Re resides – one to the left and one to the right when facing the rising Sun. These columns are delicately carved, and it is the inscriptions that suggest the sunrise alignment is deliberate. The left-hand column has an image of the Pharaoh embracing Amun-Re, and on one face are three carved papyrus stems – a plant that only grows along the northern reaches of the Nile. The right-hand column is similar in design, except the Pharaoh embraces Amun-Re wearing the crown of upper Egypt, which is south of Karnak. The three carved stems on this column are lotus blossoms, which only grow to the south.

It seems clear therefore that the columns are positioned and decorated to mark the compass directions around the temple, which is persuasive evidence that the heart of this building is aligned to capture the light from an important celestial event – the rising of the Sun in midwinter. It is a colossal representation of the details of our planet's orientation and orbit around our nearby star.

The temple represents the fascination of the ancient Egyptians with the movement of the lights they saw in the sky. Their instinct to venerate them was pre-scientific, but the building also appears to enshrine a deepening awareness of the geometry of the cosmos. By observing the varying position of sunrise, an understanding of the Earth's cycles and seasons developed, which provided essential information for planting and harvesting crops at optimum times. The development of more advanced agricultural techniques made civilisations more prosperous, ultimately giving them more time for thought, philosophy, mathematics and science. So astronomy began a virtuous cycle through which the quest to understand the heavens and their meaning lead to practical and intellectual riches beyond the imagination of the ancients.

The step from observing the regularity in the movement of the heavenly lights to modern science took much of recorded human history. The ancient Greeks began the work, but the correct description of the motion of the Sun, Moon and planets across the sky was discovered in the seventeenth century by Johannes Kepler. Removing the veil of the divine to reveal the true beauty of the cosmos was a difficult process, but the rewards that stem from that innate human fascination with the lights in the sky have proved to be incalculable ◉

By following the light we have mapped our place among the hundreds of billions of stars that make up the Milky Way Galaxy. We have visited our nearest star, Proxima Centauri, and measured its chemical compositions, and those of thousands of other stars in the sky. We have even journeyed deep into the Milky Way and stared into the black hole that lies at the centre of our galactic home. But this is just the beginning...

RIGHT: The Universe is an awe-inspiring place, full of wonder and demanding the answers to so many questions. We have so much to learn and so many places to explore.

OUR PLACE IN THE UNIVERSE

The scale of the Universe is almost impossible to comprehend and yet that's exactly what we've been able to do from the vantage point of the small rock we call Earth. As we have discovered the grand cycles that play out above our heads we have come to realise that we are part of a structure that extends way beyond our solar system and the 200 billion stars that make up our galaxy.

SOLAR SYSTEM

Mercury
Venus
Sun
Uranus
Saturn
Mars
Earth
Jupiter
Neptune

Lalande 21185
Ross 128
Wolf 359
Barnard's Star
Procyon A, B
Proxima Centauri
61 Cygni A, B
Luyten's Star
Groombridge 34 A, B
Alpha Centauri A, B
Sirius A, B
SOLAR SYSTEM
Ross 154
Epsilon Eridani
Ross 248
Tau Ceti
Lacaille 9352
Epsilon Indi

SUN'S NEIGHBOURS
20 LIGHT-YEARS

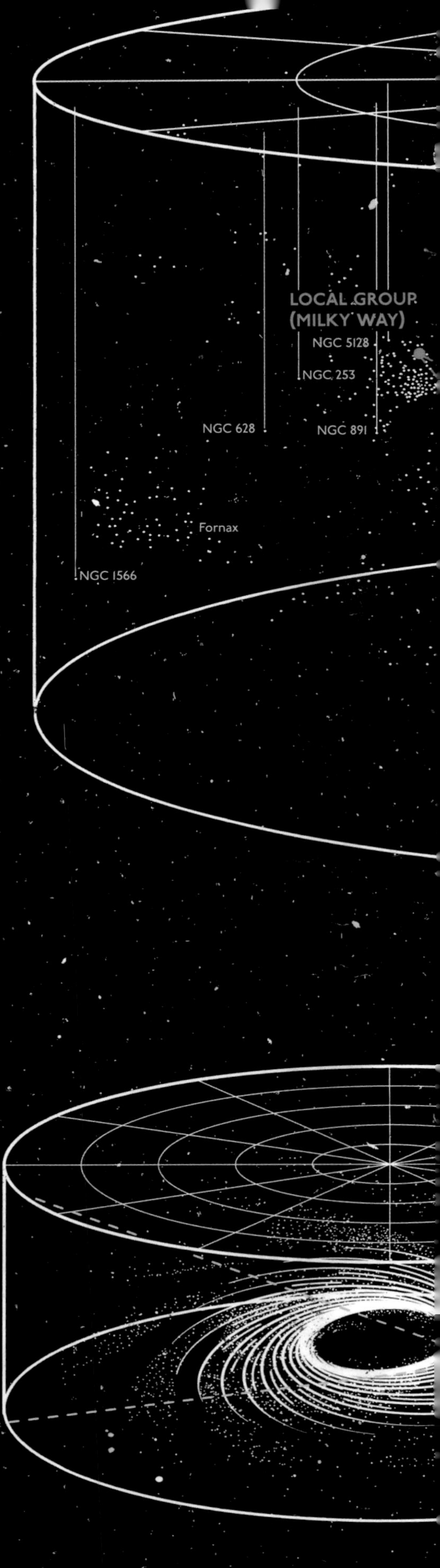

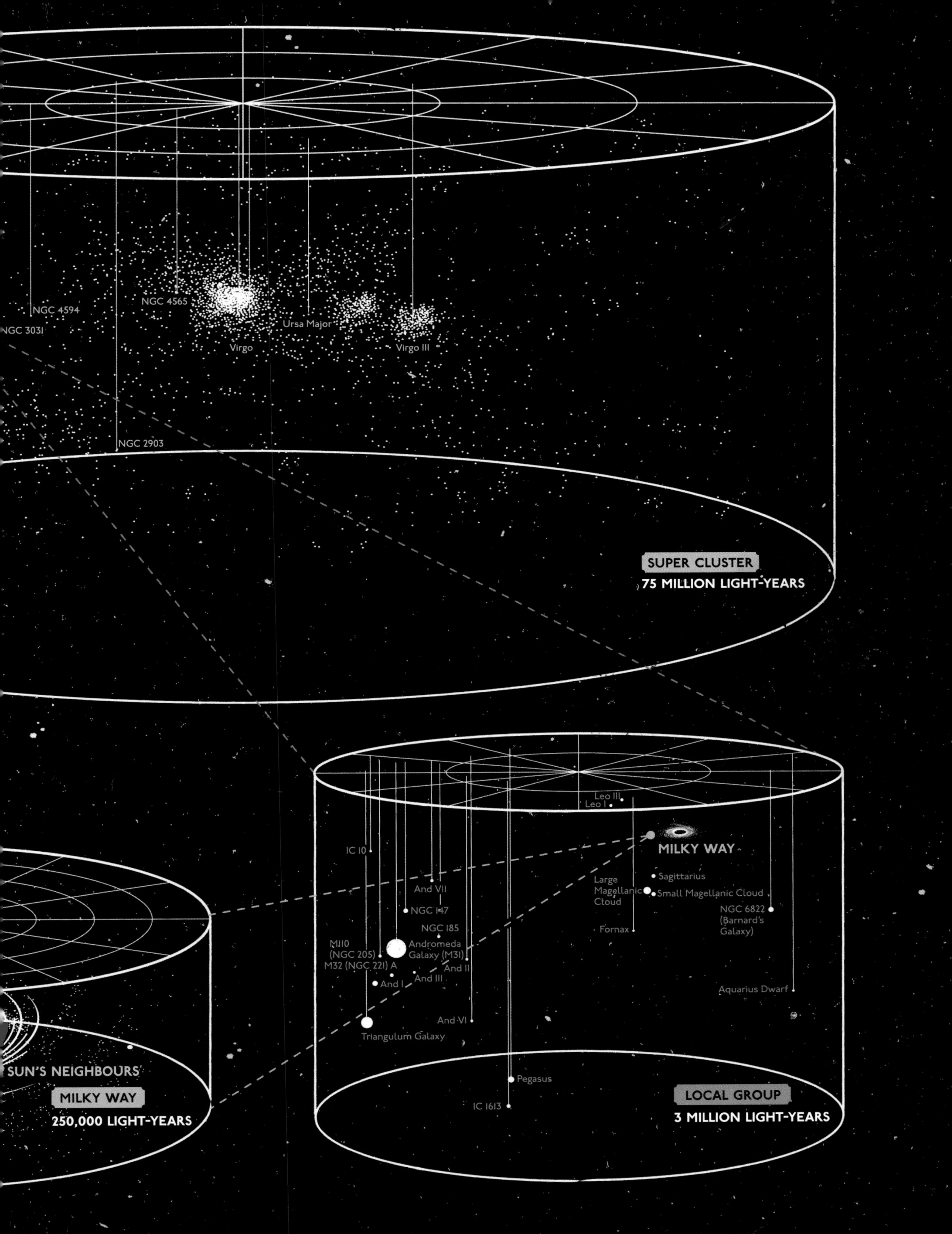
NGC 4594
NGC 3031
NGC 4565
Virgo
Ursa Major
Virgo III
NGC 2903
SUPER CLUSTER
75 MILLION LIGHT-YEARS
Leo III
Leo I
MILKY WAY
IC 10
Sagittarius
Large Magellanic Cloud
Small Magellanic Cloud
And VII
NGC 147
NGC 6822 (Barnard's Galaxy)
NGC 185
Fornax
M110 (NGC 205)
Andromeda Galaxy (M31)
M32 (NGC 221) A
And II
And III
And I
Aquarius Dwarf
And VI
Triangulum Galaxy
SUN'S NEIGHBOURS
MILKY WAY
250,000 LIGHT-YEARS
Pegasus
LOCAL GROUP
IC 1613
3 MILLION LIGHT-YEARS

OUR GALACTIC NEIGHBOURHOOD

From our small rock, we have a grandstand seat to explore our local galactic neighbourhood. Our nearest star, the Sun, is 150 million kilometres (93 million miles) away, but each night when this star disappears from view, thousands more fill the night sky. In the most privileged places on Earth, up to 10,000 stars can be seen with the naked eye, and all of them are part of the galaxy we call home.

A galaxy is a massive collection of stars, gas and dust bound together by gravity. It is a place where stars live and die, where the life cycles of our universe are played out on a gargantuan scale. We think there are around 100 billion galaxies in the observable universe, each containing many millions of stars. The smallest galaxies, known as dwarf galaxies, have as few as ten million stars. The biggest, the giants, have been estimated to contain in the region of 100 trillion. It is now widely accepted that galaxies also contain much more than just the matter we can see using our telescopes. They are thought to have giant halos of dark matter, a new form of matter unlike anything we have discovered on Earth and which interacts only weakly with normal matter. Despite this, its gravitational effect dominates the behaviour of galaxies today and most likely dominated the formation of the galaxies in the early Universe. This is because we now think that around 95 per cent of the mass of galaxies such as our own Milky Way is made up of dark matter. In some sense this makes the luminous stars, planets, gas and dust an after-thought, although because it is highly unlikely that dark matter can form into complex and beautiful structures like stars, planets and people, one might legitimately claim that it's rather less interesting. The search for the nature of dark matter is one of the great challenges for twenty-first-century physics. We shall return to the fascinating subject of dark matter later in the book.

The word 'galaxy' comes from the Greek word *galaxias*, meaning milky circle. It was first used to describe the galaxy that dominates our night skies, even though the Greeks could have had no concept of its true scale. Watching the core of our galaxy rise in the night sky is one of nature's greatest spectacles, although regrettably the light of our cities has robbed us of this majestic nightly display. For many people it looks like the rising of storm clouds on the horizon, but as the Earth turns nightly towards the centre of our galaxy, the hazy band of light reveals itself as clouds of stars – billions of them stretching thousands of light years inwards towards the galactic centre. In Greek mythology this ethereal light was described as the spilt milk from the breast of Zeus's wife, Hera, creating a faint band across the night sky. This story is the origin of the modern name for our galaxy – the Milky Way. The name entered the English language not from a scientist, but from the pen of the Medieval poet, Geoffrey Chaucer: 'See yonder, lo, the Galaxyë, Which men clepeth

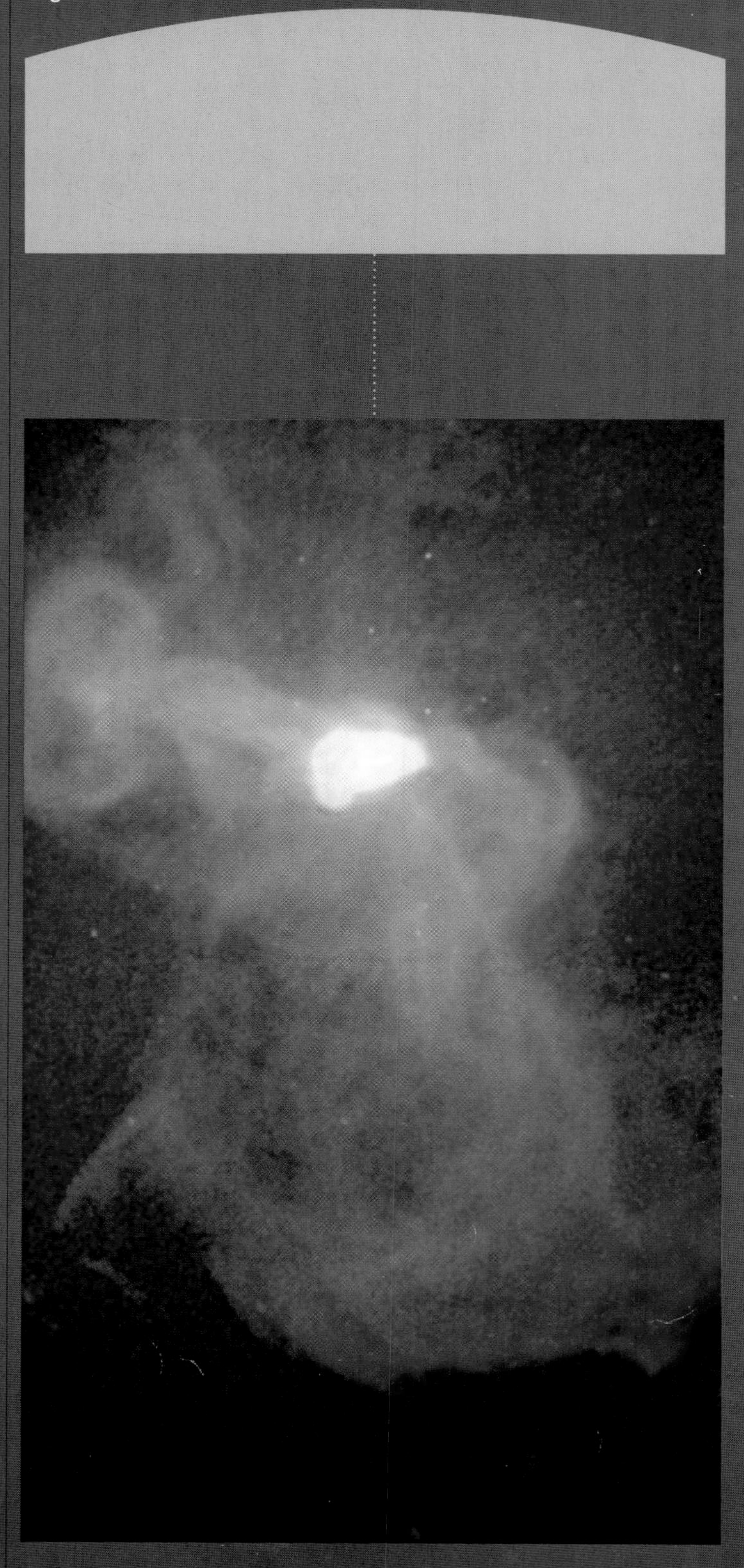

ABOVE: M87, also known as Virgo A and Messier 87, is a giant elliptical galaxy located 54 million light years away from Earth in the Virgo Cluster. In this image the central jet is visible, which is a powerful beam of hot gas produced by a massive black

M31
Andromeda Galaxy

Milky Way Galaxy

M51
Whirlpool Galaxy

M33
Triangulum Galaxy

NUBECULA MINOR
Small Magellanic cloud

DWARF GALAXY
Zwicky 18

ABOVE: Taken in December 2010, this is the most detailed picture of the Andromeda Galaxy, or M31, taken so far. It is our largest and closest spiral galaxy, and in this picture we can clearly see rings of new star formations developing.

TOP: This image of the galaxy M51 clearly shows how it got its other name: the Whirlpool Galaxy. The spiral shape of the galaxy is immediately obvious, with curving arms of pinky-red, star-forming regions and blue star clusters.

ABOVE: Zwicky 18 was once thought to be the youngest galaxy, as its bright stars suggested it was only 500 million years old. However, recent Hubble Space Telescope images have identified older stars within it, making the galaxy as old as others but with new star formations.

TOP: M33, also known as the Triangulum, or Pinwheel, Galaxy is the third-largest in the Local Group of galaxies after the Milky Way and Andromeda Galaxies, of which it is thought to be a satellite.

MAPPING THE MILKY WAY GALAXY

Our galaxy, the Milky Way, contains somewhere between 200 and 400 billion stars, depending on the number of faint dwarf stars that are difficult for us to detect. The majority of stars lie in a disc around 100,000 light years in diameter and, on average, around 1,000 light years thick. These vast distances are very difficult to visualise. A distance of 100,000 light years means that light itself, travelling at 300,000 kilometres (186,000 miles) per second, would take 100,000 years to make a journey across our galaxy. Or, to put it another way, the distance between the Sun and the outermost planet of our solar system, Neptune, is around four light hours – that's one-sixth of a light day. You would have to lay around 220 million solar systems end to end to cross our galaxy.

At the centre of our galaxy, and possibly every galaxy in the Universe, there is believed to be a super-massive black hole. Astronomers believe this because of precise measurements of the orbit of a star known as S2. This star orbits around the intense source of radio waves known as Sagittarius A* (pronounced 'Sagittarius A-star') that sits at the galactic centre. S2's orbital period is just over fifteen years, which makes it the fastest-known orbiting object, reaching speeds of up to 2 per cent of the speed of light. If the precise orbital path of an object is known, the mass of the thing it is orbiting around can be calculated, and the mass of Sagittarius A* is enormous, at 4.1 million times the mass of our sun. Since the star S2 has a closest approach to the object of only seventeen light hours, it is known that Saggitarus A* must be smaller than this, otherwise S2 would literally bump into it. The only known way of cramming 4.1 million times the mass of the Sun into a space less than 17 light hours across is as a black hole, which is why astronomers are so confident that a giant black hole sits at the centre of the Milky Way. These observations have recently been confirmed and refined by studying a further twenty-seven stars, known as the S-stars, all with orbits taking them very close to Sagittarius A*.

Beyond the S-stars, the galactic centre is a melting pot of celestial activity, filled with all sorts of different systems that interact and influence each other. The Arches Cluster

The distance between the Sun and the outermost planet of our solar system, Neptune, is around four light hours – that's one-sixth of a light day. You would have to lay around 220 million solar systems end to end to cross our galaxy.

LEFT: This artist's impression shows the Arches Cluster, the densest known cluster of young stars in the Milky Way Galaxy.

ABOVE LEFT: Along with the Arches Cluster, the Quintuplet Cluster is located near the centre of the Milky Way Galaxy.

ABOVE RIGHT: The bright white dot in the centre of this image is the Pistol Star, one of the brightest stars in our galaxy.

is the densest known star cluster in the galaxy. Formed from about 150 young, intensely hot stars that dwarf our sun in size, these stars burn brightly and are consequently very short-lived, exhausting their supply of hydrogen in just a couple of million years. The Quintuplet Cluster contains one of the most luminous stars in our galaxy, the Pistol Star, which is thought to be near the end of its life and on the verge of becoming a supernova (see pages 130–1). It is in central clusters like the Arches and the Quintuplet that the greatest density of stars in our galaxy can be found. As we move out from the crowded galactic centre, the number of stars drops with distance, until we reach the sparse cloud of gas in the outer reaches of the Milky Way known as the Galactic Halo.

In 2007, scientists using the Very Large Telescope (VLT) at the Paranal Observatory in Chile were able to observe a star in the Galactic Halo that is thought to be the oldest object in the Milky Way. HE 1523-0901 is a star in the last stages of its life; known as a red giant, it is a vast structure far bigger than our sun, but much cooler at its surface. HE 1523-0901 is interesting because astronomers have been able to measure the precise quantities of five radioactive elements – uranium, thorium, europium, osmium and iridium – in the star. Using a technique very similar to carbon dating (a method archaeologists use to measure the age of organic material on Earth), astronomers have been able to get a precise age for this ancient star. Radioactive dating is an extremely precise and reliable technique when there are multiple 'radioactive clocks' ticking away at once. This is why the detection of five radioactive elements in the light from HE 1523-0901 was so important. This dying star turns out to be 13.2 billion years old – that's almost as old as the Universe itself, which began just over 13.7 billion years ago. The radioactive elements in this star would have been created in the death throes of the first generation of stars, which ended their lives in supernova explosions in the first half a billion years of the life of the Universe (see Chapter 2) ◉

THE SHAPE OF OUR GALAXY

As well as being vast and very, very old, our galaxy is also beautifully structured. Known as a barred spiral galaxy, it consists of a bar-shaped core surrounded by a disc of gas, dust and stars that creates individual spiral arms twisting out from the centre. Until very recently, it was thought that our galaxy contained only four spiral arms – Perseus, Norma, Scutum–Centaurus and Carina–Sagittarius, with our sun in an offshoot of the latter called the Orion spur – but there is now thought to be an additional arm, called the Outer arm, an extension to the Norma arm.

Close to the inner rim of the Orion spur is the most familiar star in our galaxy. The Sun was once thought to be an average star, but we now know that it shines brighter than 95 per cent of all other stars in the Milky Way. It's known as a main sequence star because it gets all its energy and produces all its light through the fusion of hydrogen into helium. Every second, the Sun burns 600 million tonnes of hydrogen in its core, producing 596 million tonnes of helium in the fusion reaction. The missing four million tonnes of mass emerges as energy, which slowly travels to the Sun's photosphere, where it is released into the galaxy and across the Universe as light

THREE MAIN TYPES OF GALAXY: ELLIPTICAL, SPIRAL AND BARRED SPIRAL

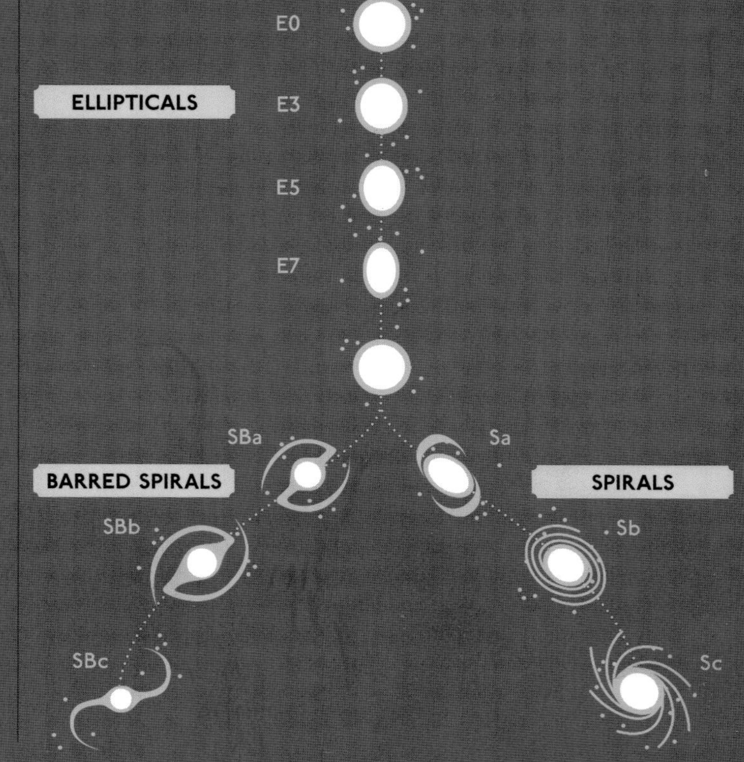

THE SPIRALLING ARMS OF THE MILKY WAY GALAXY

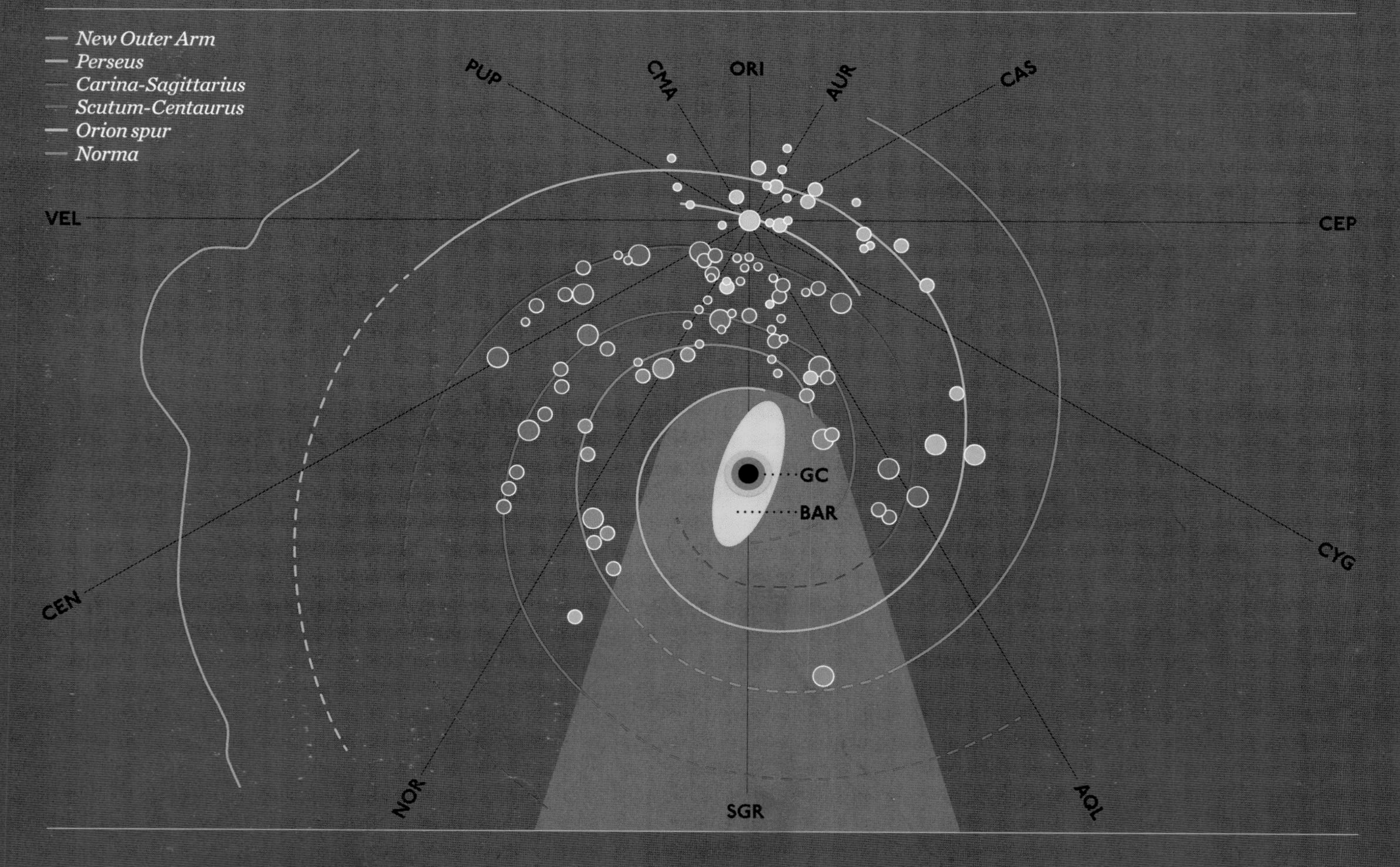

BELOW: The Andromeda Galaxy is our nearest galactic neighbour, and our own Milky Way Galaxy is believed to look very much like it.

BOTTOM: Located 5,000 light years away, the Lagoon Nebula is one of a handful of active star-forming regions in our galaxy that are visible from Earth with the naked eye.

A STAR IS BORN

Our sun is in the middle of its life cycle, but look out into the Milky Way and we can see the whole cycle of stellar life playing out. Roughly once a year a new light appears in our galaxy, as somewhere in the Milky Way a new star is born.

The Lagoon Nebula is one such star nursery; within this giant interstellar cloud of gas and dust, new stars are created. Discovered by French astronomer Guillaume Le Gentil in 1747, this is one of a handful of active star-forming regions in our galaxy that are visible with the naked eye. This huge cloud is slowly collapsing under its own gravity, but slightly denser regions gradually accrete more and more matter, and over time these clumps grow massive enough to turn into stars.

The centre of this vast stellar nursery, known as the Hourglass, is illuminated by an intriguing object known as Herschel 36. This star is thought to be a 'ZAMS' star (zero ago main sequence) because it has just begun to produce the dominant part of its energy from hydrogen fusion in its core. Recent measurements suggest that Herschel 36 may actually be three large young stars orbiting around each other, with the entire system having a combined mass of over fifty times that of our sun. This makes Herschel 36 a true system of giants.

Eventually Herschel 36 and all the stars in the Milky Way will die, and when they do, many will go out in a blaze of glory.

Eta Carinae is a pair of billowing gas and dust clouds that are the remnants of a stellar explosion from an unstable star system. The system consists of at least two giant stars, and shines with a brightness four million times that of our sun. One of these stars is thought to be a Wolf-Rayet star. These stars are immense, over twenty times the mass of our sun, and are engaged in a constant struggle to hang onto their outer layers, losing vast amounts of mass every second in a powerful solar wind. In 1843, Eta Carinae became one of the brightest stars in the Universe when it exploded. The blast spat matter out at nearly 2.5 million kilometres (1.5 million miles) an hour, and was so bright that it was thought to be a supernova explosion. Eta Carinae survived intact and remains buried deep inside these clouds, but its days are numbered. Because of its immense mass, the Wolf-Rayet star is using up its hydrogen fuel at a ferocious rate. Within a few hundred thousand years, it is expected that the star will

Out in the Milky Way we can see the whole cycle of stellar life playing out. Roughly once a year a new light appears, as somewhere in the Milky Way a new star is born.

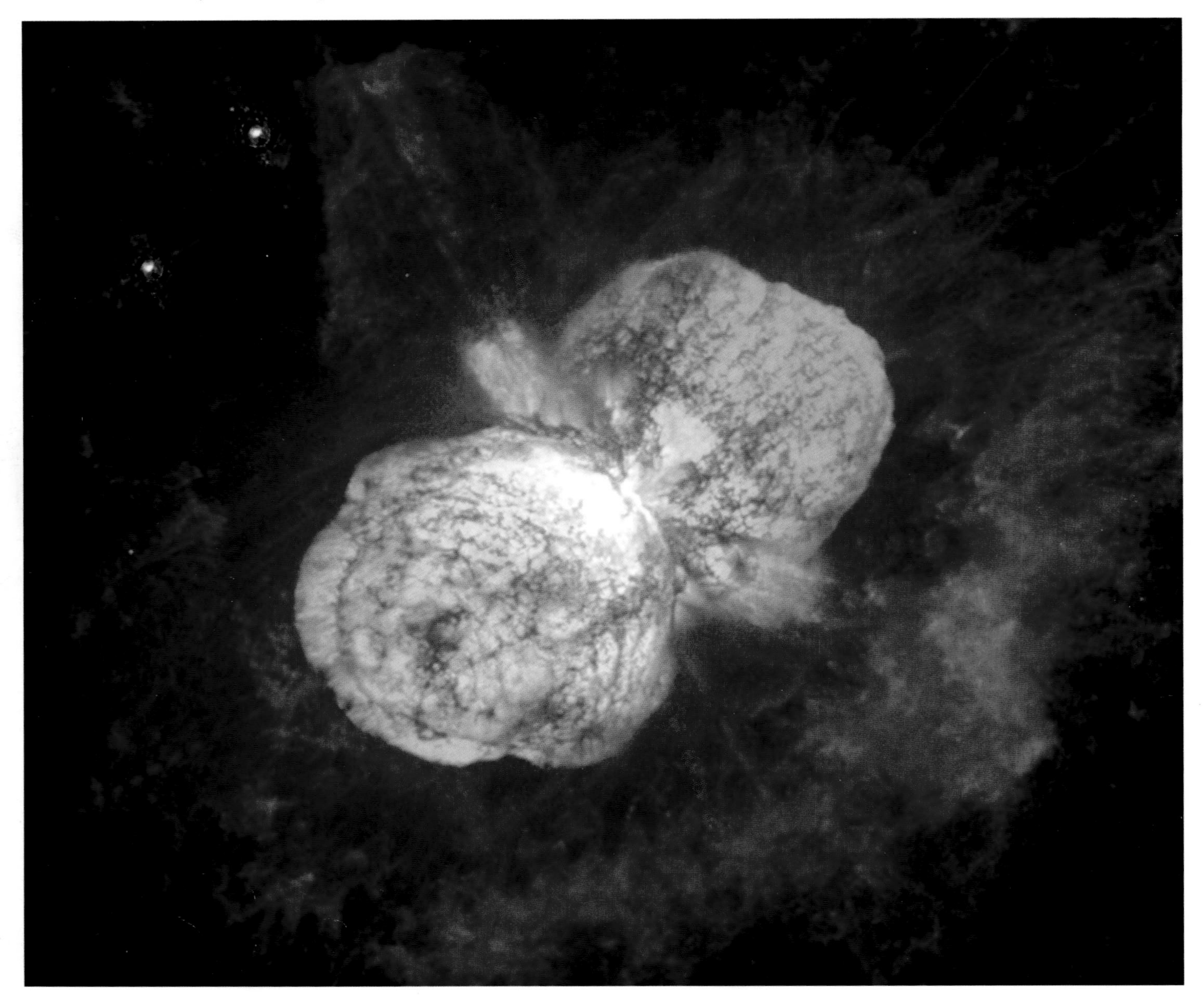

explode in a supernova or even a hypernova (the biggest explosion in the known Universe), although its fate may be sealed a lot sooner. In 2004, an explosion thought to be similar to the 1843 Eta Carinae event was seen in a galaxy over seventy million light years from the Milky Way. Just two years later, the star exploded as a supernova. Eta Carinae is very much closer – at a distance of only 7,500 light years – so as a supernova it may shine so brightly that it will be visible from Earth even in daylight.

Seeing the light from these distant worlds and watching the life cycle of the Universe unfold is a breathtaking reminder that light is the ultimate messenger; carrying information about the wonders of the Universe to us across interstellar and intergalactic distances. But light does much more than just allow us to see these distant worlds; it allows us to journey back through time, providing a direct and real connection with our past. This seemingly impossible state of affairs is made possible not only because of the information carried by the light, but by the properties of light itself ◉

LEFT: Eta Carinae is one of the most massive and visible stars in the night sky, but because of its mass it is also the most volatile and most likely to explode in the near future.

BELOW: Eventually all the stars in the Milky Way will die, many in spectacular explosions. Herschel 36 was formed from just such a stellar explosion, which occurred within the Eta Carinae system.

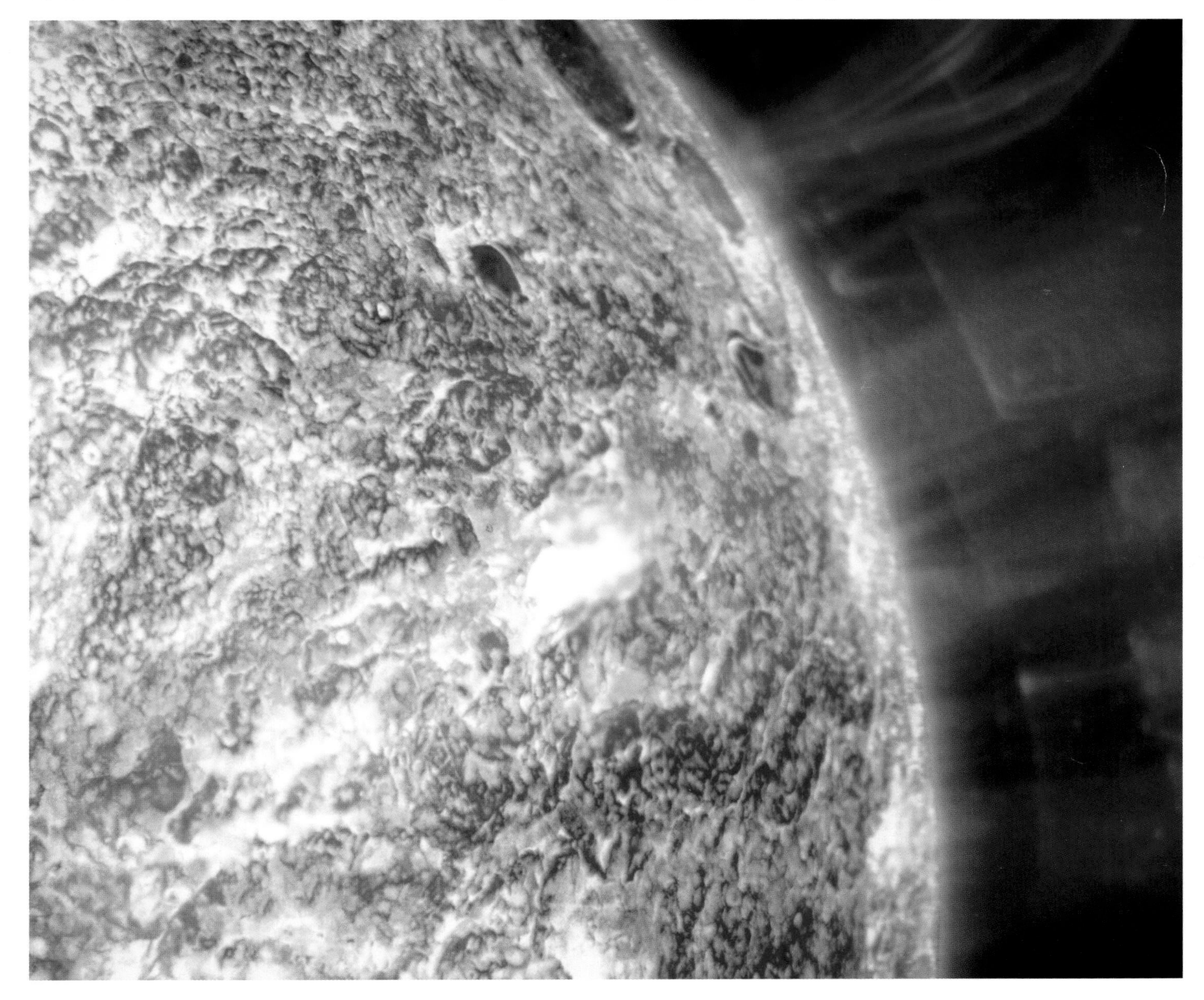

WHAT IS LIGHT?

If we aspire to understand the world around us, one of the most basic questions we must ask is about the nature of light. It is the primary way in which we observe our own planet, and the only way we will ever be able to explore the Universe beyond our galaxy. For now, even the stars are far beyond our reach, and we rely on their light alone for information about them. By the seventeenth century, many renowned scientists were studying the properties of light in detail, and parallel advances in engineering and science both provided deep insights and catalysed each other. The studies of Kepler, Galileo and Descartes, and some of the later true greats of physics – Huygens, Hooke and Newton – were all fuelled by the desire to build better lenses for microscopes and telescopes to enable them to explore the Universe on every scale, and to make great scientific discoveries and advances in the basic science itself.

YOUNG'S DOUBLE-SLIT EXPERIMENT

By the end of the seventeenth century, two competing theories for light had emerged – both of which are correct. On one side was Sir Isaac Newton, who believed that light was composed of particles – or 'corpuscles', as he called them in his *Hypothesis of Light*, published in 1675. On the other were Newton's great scientific adversary, Robert Hooke, and the Dutch physicist and astronomer, Christiaan Huygens. The particle/wave debate rumbled on until the turn of the nineteenth century, with most physicists siding with Newton. There were some notable exceptions, including the great mathematician Leonhard Euler, who felt that the phenomena of diffraction could only be explained by a wave theory. In 1801, the English doctor Thomas Young appeared to settle the matter once and for all when he reported the results from his famous double-slit experiment, which clearly showed that light diffracted, and therefore must travel in the form of a wave.

Diffraction is a fascinating and beautiful phenomena that is very difficult to explain without waves. If you shine light onto a screen through a barrier with a very thin slit cut into it, you don't see a bright light on the screen opposite the slit, but instead you see a complex but regular pattern of light and dark areas.

The explanation for this is that when you mix lots of waves together they don't only have to add up. Imagine two waves on top of each other with exactly the same wavelength and wave height (technically known as the amplitude), but aligned precisely so that the peak of one wave lies directly on the trough of the other (in more technical language, we say that the waves are 180 degrees out of phase), and so the waves cancel each other out. If these waves were light waves you would get darkness! This is exactly what is seen in diffraction experiments through small slits. The slits act like lots of little sources of light, all slightly displaced from one another. This means that there will be places beyond the slits where the waves cancel each other out, and places where they will add up, leading to the light and dark areas seen by experimenters like Young. This was taken as clear evidence that light was some kind of wave – but waves of what?

BELOW: The results of Young's double-slit experiment are revealed in this detailed, wide pattern. The experiment demonstrates the inseparability of the wave and particle natures of light and other quantum particles.

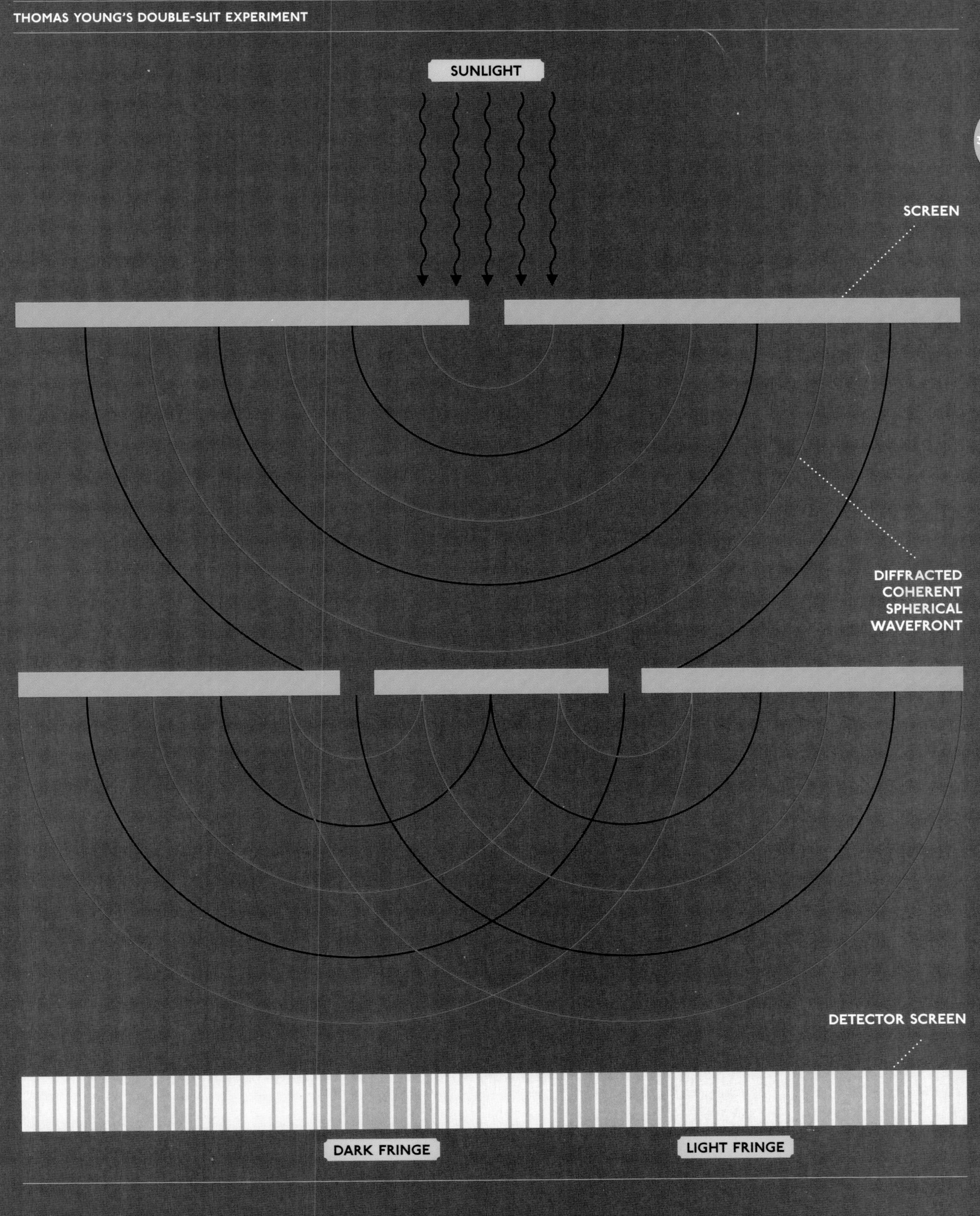
SUNLIGHT
SCREEN
DIFFRACTED COHERENT SPHERICAL WAVEFRONT
DETECTOR SCREEN
DARK FRINGE
LIGHT FRINGE

MESSENGERS FROM ACROSS THE OCEAN OF SPACE

As is often the way in science, the correct explanation for the nature of light came from an unlikely source. In the mid-nineteenth century, the study of electricity and magnetism engaged many great scientific minds. At the Royal Institution in London, Michael Faraday was busy doing what scientists do best – playing around with wire and magnets. He discovered that if you push a magnet through a coil of wire, an electric current flows through the wire while the magnet is moving. This is a generator; the thing that sits in all power stations around the world today, providing us with electricity. Faraday wasn't interested in inventing the foundation of the modern world, he just wanted to learn about electricity and magnetism. He encoded his experimental findings in mathematical form – known today as Faraday's Law of Electromagnetic Induction. At around the same time, the French physicist and mathematician André-Marie Ampère discovered that two parallel wires carrying electric currents experience a force between them; this force is still used today to define the ampere, or amp – the unit of electric current. A single amp is defined as the current that must flow along two parallel wires of infinite length and negligible diameter to produce an attractive force of 0.0000007 Newtons between them. Next time you change a thirteen-amp fuse in your plug, you are paying a little tribute to the work of Ampère. Today, the mathematical form of this law is called Ampère's Law.

By 1860, a great deal was known about electricity and magnetism. Magnets could be used to make electric currents flow, and flowing electric currents could deflect compass needles in the same way that magnets could. There was clearly a link between these two phenomena, but nobody had come up with a unified description. The breakthrough was made by the Scottish physicist James Clerk Maxwell, who, in a series of papers in 1861 and 1862, developed a single theory of electricity and magnetism that was able to explain all of the experimental work of Faraday, Ampère and others. But Maxwell's crowning glory came in 1864, when he published a paper that is undoubtedly one of the greatest achievements in the history of science. Albert Einstein later described Maxwell's 1860s papers as 'the most profound and the most fruitful that physics has experienced since the time of Newton.' Maxwell discovered that by unifying electrical and magnetic phenomena together into a single mathematical theory, a startling prediction emerges.

Electricity and magnetism can be unified by introducing two new concepts: electric and magnetic fields. The idea of a field is central to modern physics; a simple example of something that can be represented by a field is the temperature in a room. If you could measure the temperature at each point in the room and note it down, eventually you would have a vast array of numbers that described how the temperature changes from the door to the windows and from the floor to the ceiling. This array of numbers is called the temperature field. In a similar way, you could introduce the concept of a magnetic field by holding a compass at places around a wire carrying an electric current and noting down how much the needle deflects, and in what direction. The numbers and directions are the magnetic field. This might

LEFT: The movement of waves across the ocean can be explained by a set of equations; Maxwell discovered a similar form of equation explained waves within magnetic fields.

BELOW: These picture strips illustrate maps of the Milky Way Galaxy as they appear in different wavelength regions.

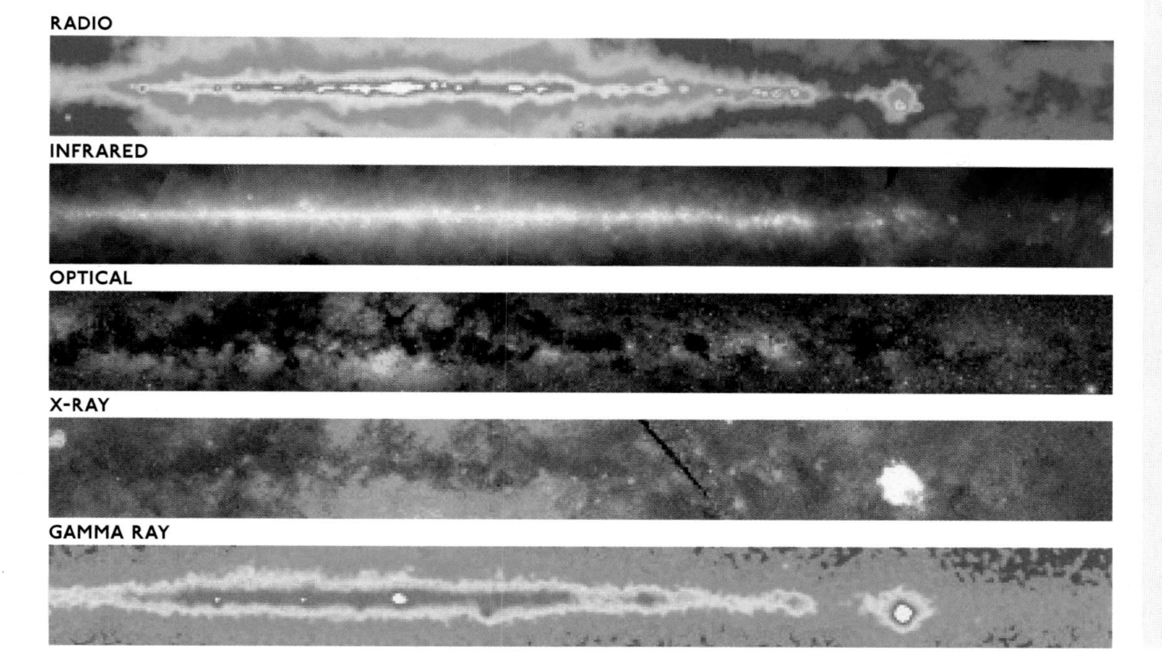

THE RELATIONSHIP BETWEEN ELECTRICITY, MAGNETISM AND THE SPEED OF LIGHT IS SUMMARIZED IN THE EQUATION:

Where c is the speed of light and the quantities μ_0 and ε_0 are related to the strengths of electric and magnetic fields. The fact that the velocity of light can be measured experimentally on a bench top with wires and magnets was the key piece of evidence that light is an electromagnetic wave.

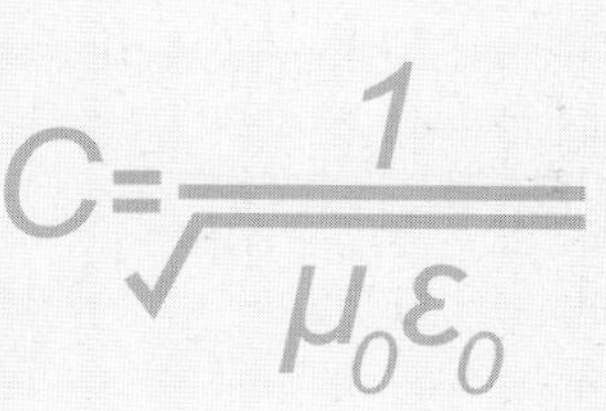

$$C = \frac{1}{\sqrt{\mu_0 \varepsilon_0}}$$

Maxwell's equations had exactly the same form as the equations that describe how soundwaves move through air or how water waves move through the ocean.

seem rather abstract and not much of a simplification, but Maxwell found that by introducing the electric and magnetic fields and placing them centre stage, he was able to write down a single set of equations that described all the known electrical and magnetic phenomena.

At this point you may be wondering what all this has to do with the story of light. Well, here is something profound that provides a glimpse into the true power and beauty of modern physics. In writing down his laws of electricity and magnetism using fields, Maxwell noticed that by using a bit of simple mathematics, he could rearrange his equations into a more compact and magically revealing form. His new equations took the form of what are known as wave equations. In other words, they had exactly the same form as the equations that describe how soundwaves move through air or how water waves move through the ocean. But waves of what? The waves Maxwell discovered were waves in the electric and magnetic fields themselves. His equations showed that as an electric field changes, it creates a changing magnetic field. But in turn as the magnetic field changes, it creates a changing electric field, which creates a changing magnetic field, and so on. In other words, once you've wiggled a few electric charges around to create a changing electric and magnetic field, you can take the charges away and the fields will continue sloshing around – as one falls, the other will rise. And this will continue to happen forever, as long as you do nothing to them.

This is profound in itself, but there is an extra, more profound conclusion. Maxwell's equations also predict exactly how fast these waves must fly away from the electric charges that create them. The speed of the waves is the ratio of the strengths of the electric and magnetic fields – quantities that had been measured by Faraday, Ampère and others and were well known to Maxwell. When Maxwell did the sums, he must have fallen off his chair. He found that his equations predicted that the waves in the electric and magnetic fields travelled at the speed of light! In other words, Maxwell had discovered that light is nothing more than oscillating electric and magnetic fields, sloshing back and forth and propelling each other through space as they do so. How beautiful that the work of Faraday, Ampère and others with coils of wire and pieces of magnets could lead to such a profound conclusion through the use of a bit of mathematics and a sprinkling of Scottish genius! In modern language, we would say that light is an electromagnetic wave.

In order to have his epiphany, Maxwell needed to know exactly what the speed of light was. Remarkably, the fact that light travels very fast, but not infinitely so, had already been known for almost two hundred years. As we will discover now, it had first been measured by Ole Romer in 1676 ◉

CHASING THE SPEED OF LIGHT

Open your eyes and the world floods in; light seems to jump from object to retina, forming a picture of the world instantaneously. Light seems to travel infinitely fast, so it is no surprise that Aristotle and many other philosophers and scientists believed light travelled 'without movement'. However, as the Greek philosophers gave more thought to the nature of light, a debate about its speed of travel ensued that continued for thousands of years.

In one corner sat eminent names such as Euclid, Kepler and Descartes, who all sided with Aristotle in believing that light travelled infinitely fast. In the other, Empedocles and Galileo, separated by almost two millennia, felt that light must travel at a finite, if extremely high, velocity. Empedocles's reasoning was elegant, pre-dating Aristotle by a century. He considered light travelling across the vast distance from the Sun to Earth, and noted that everything that travels must move from one point to another. In other words, the light must be somewhere in the space between the Sun and the Earth after it leaves the Sun and before it reaches the Earth. This means it must travel with a finite velocity. Aristotle dismissed this argument by invoking his idea that light is simply a presence, not something that moves between things. Without experimental evidence, it is impossible to decide between these positions simply by thinking about it!

Galileo set out to measure the speed of light using two lamps. He held one and sent an assistant a large distance away with another. When they were in position, Galileo opened a shutter on his lamp, letting the light out. When his assistant saw the flash, he opened his shutter, and Galileo attempted to note down the time delay between the opening of his shutter and his observation of the flash from his assistant's lamp. His conclusion was that light must travel extremely rapidly, because he was unable to determine its speed. Galileo was, however, able to put a 'limit' on the speed of light, noting that it must be at least ten times faster than the speed of sound. He was able to do this because if it had been slower, he should have been able to measure a time delay. So, the inability to measure the speed of light was not deemed a 'no result', but in fact revealed that light travels faster than his experiment could quantify.

The first experimental determination that the speed of light was not infinite was made by the seventeenth-century

BELOW: The question, how fast is the speed of light, has plagued scientists for thousands of years. Part of the answer came from observing how light travels between points: from the Sun to Earth.

BELOW: These spectacular star trails are produced in the sky as a result of diurnal motion. This is the motion created as Earth spins on its axis at fifteen degrees per hour, rotating once over twenty-four hours.

Danish astronomer, Ole Romer. In 1676, Romer was attempting to solve one of the great scientific and engineering challenges of the age; telling the time at sea. Finding an accurate clock was essential to enable sailors to navigate safely across the oceans, but mechanical clocks based on pendulums or springs were not good at being bounced around on the ocean waves and soon drifted out of sync. In order to pinpoint your position on Earth you need the latitude and longitude. Latitude is easy; in the Northern Hemisphere, the angle of the North Star (Polaris) above the horizon is your latitude. In the Southern Hemisphere, things are more complicated because there is no star directly over the South Pole, but it is still possible with a little astronomical know-how and trigonometry to determine your latitude with sufficient accuracy for safe navigation.

Longitude is far more difficult because you can't just determine it by looking at the stars; you have to know which time zone you are in. Greenwich in London is defined as zero degrees longitude; as you travel west from Greenwich across the Atlantic, your time zone shifts so that in New York it's earlier in the day than in London. Conversely, as you travel east from Greenwich your time zone shifts so that in Moscow or Tokyo it's later in the day than in London.

Your precise time zone at any point on Earth's surface is defined by the point at which the Sun crosses an imaginary arc across the sky between the north and south points on your horizon, passing through the celestial pole (the point marked by the North Star in the Northern Hemisphere). Astronomers call this arc the Meridian. The point at which the Sun crosses the Meridian is also the point at which it reaches its highest position in the sky on any given day as it journeys from sunrise in the east to sunset in the west. We call this time noon, or midday. Earth rotates once on its axis every twenty-four hours – fifteen degrees every hour. This means two points on Earth's surface that are separated by fifteen degrees of longitude will measure noon exactly one hour apart. So to determine your longitude, set a clock to read 12 o'clock when the Sun reaches the highest point in the sky at Greenwich. If it reads 2pm when the Sun reaches its highest point in the sky where you are, you are thirty degrees to the west of Greenwich. Easy, except that you need a very accurate clock that keeps time for weeks or months on end ◉

THE SEARCH FOR A COSMIC CLOCK

RIGHT: Jupiter appears spotty in this false-colour picture from the Hubble Space Telescope's near-infrared camera. The three black spots are the shadows of the moons Ganymede (top left), Io (left) and Callisto. The white spot above centre is Io, while the blue spot (upper right) is Ganymede. Callisto is out of the image to the right.

In the early seventeenth century, King Philip III of Spain offered a prize to anyone who could devise a method for precisely calculating longitude when out of sight of land. The technological challenge of building sufficiently accurate clocks was too great, so scientists began to look for high-precision natural clocks, and it seemed sensible to look to the heavens. Galileo, having discovered the moons of Jupiter, was convinced he could use the orbits of these moons as a clock, as they regularly passed in and out of the shadow of the giant planet. The principle is beautifully simple; Jupiter has four bright moons that can be seen relatively easily from Earth, and the innermost moon, Io, goes around the planet every 1.769 days, precisely. One might say that Io's orbit is as regular as clockwork, therefore by watching for its daily disappearance and re-emergence from behind Jupiter's disc you have a very accurate and unchanging natural clock. Thus by using the Jovian system as a cosmic clock, Galileo devised an accurate system for keeping time. Observing the eclipses of these tiny pinpoints of light around three-quarters of a billion kilometres (half a billion miles) from Earth from a rolling ship was impractical, however, so although the logic was sound, Galileo failed to win the King's prize. Despite this, it was clear this technique could be used to measure longitude accurately on land, where stable conditions and high-quality telescopes were available. Thus observing and cataloguing the eclipses of Jupiter's moons, particularly Io, became a valuable astronomical endeavour.

By the mid-seventeenth century, Giovanni Cassini was leading the study of Jupiter's moons. He pioneered the use of Io's eclipses for the measurement of longitude and published tables detailing on what dates the eclipses should be visible from many locations on Earth, together with high-precision predictions of the times. In the process of further refining his longitude tables, he sent one of his astronomers, Jean Picard, to the Uraniborg Observatory near Copenhagen, where Picard employed the help of a young Danish astronomer, Ole Romer. Over some months

LEFT: These sketches (published in Istoria e Dimonstrazione in 1613) show the changing position of the moons of Jupiter over 12 days. Jupiter is represented by the large circle, with the four moons as dots on either side.

ABOVE: Ole Romer's recorded observations show his detailed research into the movement of Io.

ROMER'S THEORY: predicting the emergence of io from behind jupiter, as seen from earth, is affected by the varying distance between earth and jupiter.

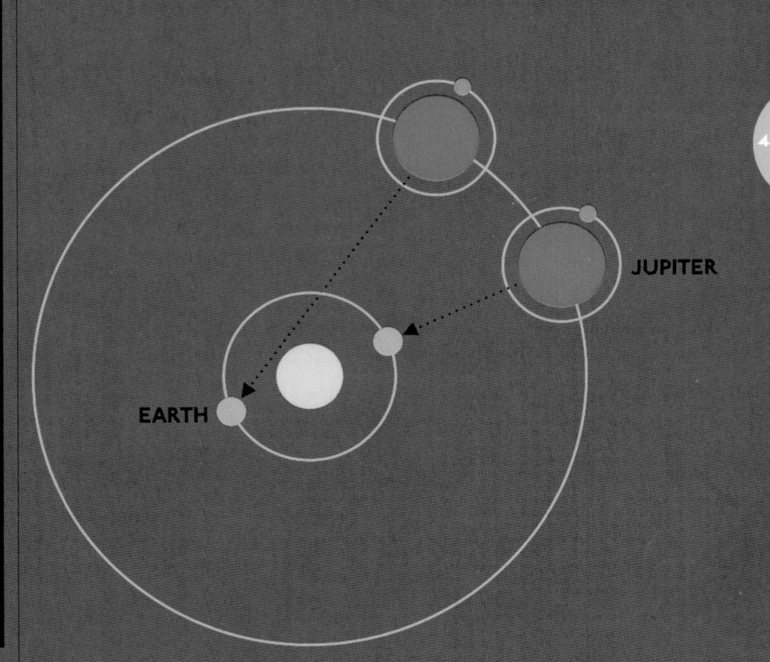

in 1671, Romer and Picard observed over one hundred of Io's eclipses, noting the times and intervals between each. He was quickly invited to work as Cassini's assistant at the Royal Observatory, where Romer made a crucial discovery. Combining the data from Uraniborg with Cassini's Paris observations, Romer noticed that the celestial precision of the Jovian clock wasn't as accurate as everyone had thought. Over the course of several months, the prediction for when Io would emerge from behind Jupiter drifted. At some times of the year there was a significant discrepancy of over twenty-two minutes between the predicted and the actual observed timings of the eclipses. This appeared to ruin the use of Io as a clock and end the idea of using it to calculate longitude. However, Romer came up with an ingenious and correct explanation of what was happening.

Romer noticed that the observed time of the eclipses drifted later relative to the predicted time as the distance between Jupiter and Earth increased as the planets orbited the Sun, then drifted back again when the distance between Jupiter and Earth began to decrease. Romer's genius was to realise that this pattern implied there was nothing wrong with the clockwork of Jupiter and Io, because the error depended on the distance between Earth and Jupiter and had nothing to do with Io itself. His explanation, which is correct, was simple. Imagine that light takes time to travel from Jupiter to Earth; as the distance between the two planets increases, so the light from Jupiter will take longer to travel between them. This means that Io will emerge from Jupiter's shadow later than predicted, simply because it takes longer for the light to reach you. Conversely, as the distance between Jupiter and Earth decreases, it takes the light less time to reach you and so you see Io emerge sooner than predicted. Factor in the time it takes light to travel between Jupiter and Earth and the theory works. Romer did this by trial and error, and was able to correctly account for the shifting times of the observed eclipses. The number that Romer actually calculated was the light travel time across the diameter of Earth's orbit around the Sun, which he found to be approximately twenty minutes. For some reason, perhaps because he felt the diameter of Earth's orbit was not known with sufficient precision, he never turned this number into the speed of light in any Earth-based units of measurement. He simply stated that it takes light twenty-two minutes to cross the diameter of Earth's orbit. The first published number for the speed of light was that obtained by the Dutch astronomer Christiaan Huygens, who had corresponded with Romer. In his 'Treatise sur la lumière' (1678), Huygens quotes a speed in strange units as 110 million toises per second. Since a toise is two metres (seven feet), this gives a speed of 220,000,000 metres per second, which is not far off the modern value of 299,792,458 metres (983,571,503 feet) per second. The error was primarily in the determination of the diameter of Earth's orbit around the Sun.

No consensus about the speed of light was reached until after Romer's death in 1710, but his correct interpretation of the wobbles in the Jovian clock still stands as a seminal achievement in the history of science. His measurement of the speed of light was the first determination of the value of what scientists call a constant of nature. These numbers, such as Newton's gravitational constant and Planck's constant, have remained fixed since the Big Bang, and are central to the properties of our universe. They are crucial in physics, and we would live (or not live, because we wouldn't exist) in a universe that was unrecognisable if their values were altered by even a tiny amount

SPEED LIMITS

Everything in our universe has a speed limit, and for much of the twentieth century humans seemed obsessed with breaking one of them. In the 1940s and 1950s the sound barrier took on an almost mythical status as engineers worldwide tried to build aircraft that could exceed the 1236 kilometres per hour (768 miles per hour) at which sound travels in air at twenty degrees Celsius. But what is the meaning of this speed limit? What is the underlying physics, and how does it affect our engineering attempts to break it?

Sound in a gas such as air is a moving disturbance of the air molecules. Imagine dropping a saucepan lid onto the floor. As it lands, it rapidly compresses the air beneath it, pushing the molecules closer together. This increases the density of the air beneath the lid, which corresponds to an increase in air pressure. In a gas, molecules will fly around to try to equalise the pressure, which is why winds develop between high and low pressure areas in our atmosphere. With a falling lid, some of the molecules in the high-pressure area beneath it will rush out to the surrounding lower-pressure areas; these increase in pressure, causing molecules to rush into the neighbouring areas, and so on. So the disturbance in the air caused by the falling lid moves outwards as a wave of pressure. The air itself doesn't flow away from the lid (this would leave an area of lower pressure around it that would have to be equalised), it is only the pulse of pressure that moves through the air.

The speed of this pressure wave is set by the properties of the air. The speed of sound in air depends on the air's temperature, which is a measure of how fast the molecules in the air are moving on average, the mass of the air molecules (air is primarily a mixture of nitrogen and oxygen) and the details of how the air responds when it is compressed (known as the 'adiabatic index'). To a reasonable approximation, the speed of the sound wave depends mainly on the average speed of the air molecules at a particular temperature.

The speed of sound is therefore not a speed limit at all; it is simply the speed at which a wave of pressure moves through the air, and there is no reason why an object shouldn't exceed this. This was known long before aircraft were invented, but it did not satisfy those who wanted to propel a human faster than sound. Many attempts were made during World War II to produce a supersonic aircraft, but the sound barrier was not breached until 14 October 1947, when Chuck Yeager became the first human to pilot a supersonic flight. Flying in the Bell–XS1, Yeager was dropped out of the bomb bay of a modified B29 bomber, through the sound barrier and into the history books.

The speed of sound is not a speed limit at all; it is simply the speed at which a wave of pressure moves through the air, and there is no reason why an object shouldn't exceed this.

ABOVE AND RIGHT: Once we reached 12,800 metres, the pilot put the Hawker Hunter into the roll and we dived down through the clouds, upside down. Almost immediately, we broke through the sound barrier.

Today, aircraft routinely break the sound barrier, but the routine element hides the fascinating aerodynamic and engineering challenges that had to be overcome so that humans could travel faster than sound. Test pilot Dave Southwood demonstrated these to me in the making of the programme in a beautiful aircraft that was not designed to break the sound barrier in level flight – the Hawker Hunter.

Designed in the 1950s, the Hawker Hunter is a legendary British jet fighter of the post-war era. Designed to fly at Mach 0.94, this aircraft cannot fly supersonic in level flight, but in the right hands it can exceed the 1,200 kilometres (745 miles) per hour to take me through the sound barrier. We climbed to 12,800 metres (42,000 feet), flipped the Hunter into an inverted dive, then plunged full-throttle towards the Bristol

Channel. In just seconds the jet smashed through the sound barrier and the air flow surrounding the jet changed, which is heard on the ground as an explosion, or a sonic boom.

So the sound barrier is not a barrier at all; it is a speed limit only for sound itself, determined by the physics of the movement of air molecules. Is the light barrier the same? It would seem from our description of light as an electromagnetic wave that is so. Why shouldn't a sufficiently powerful aircraft or spacecraft be able to fly faster than a wave in electric and magnetic fields? The answer is that the 'light barrier' is of a totally different character and cannot be smashed through, even in principle. The reason for this is that light speed plays a much deeper role in the Universe than just being the speed at which light travels. A true understanding of the role of this speed, 299,792,458 metres (983,571,503 feet) per second, was achieved in 1905 by Albert Einstein in his special theory of relativity. Einstein, inspired by Maxwell's work, wrote down a theory in which space and time are merged into a single entity known as 'spacetime'. Einstein suggested we should not see our world as having only three directions – north/south, east/west and up/down, as he added a fourth direction – past/future. Hence spacetime is referred to as four-dimensional, with time being the fourth dimension.

A full explanation of this is beyond the scope of this book, suffice to say that Einstein was forced into this bold move primarily because Maxwell's equations for electricity and magnetism were incompatible with Newton's 200-year-old laws of motion. Einstein abandoned the Newtonian ideas of space and time as separate entities and merged them. In Einstein's theory there is a special speed built into the structure of spacetime itself that everyone must agree on, irrespective of how they are moving relative to each other. This special speed is a universal constant of nature that will always be measured as precisely 299,792,458 metres (983,571,503 feet) per second, at all times and all places in the Universe, no matter what they are doing. This is critical in Einstein's theory because it stops us doing something strange in spacetime; if past/future is simply another direction like north/south, why can't we wander backwards and forwards in it? Why can we only travel into the future, not the past?

In Einstein's theory of relativity it is the existence of this unanimously agreed special speed that makes time direction different to that of space and prevents time travel. In this sense, the special speed is built into the fabric of space and time itself and plays a deep role in the structure of our universe. What does it have to do with the speed of light? Nothing much! There is a reason why light goes at this speed, and it seems to be a complete coincidence. In Einstein's theory, anything that has no mass is compelled to travel at the special speed through space. Conversely, anything that has mass is compelled to travel slower than this speed. Particles of light, photons, have no mass, so they travel at the speed of light. There is no deep reason we know of why photons have to be massless particles, so no deep reason why light travels at the speed of light! We only call the special speed 'light speed' because it was discovered by measuring the speed of light.

The key point is that the speed of light is a fundamental property of the Universe because it is built into the fabric of space and time itself. Travelling faster than this speed is impossible, and even travelling at it is impossible if you have mass. It is this property of the Universe that protects the past from the future and prevents time travel into the past ◉

TIME TRAVEL

Without realising it, we are all travelling back in time by the most miniscule amount. The consequence of light travelling fast, but not infinitely fast, is that you see everything as it was in the past. In everyday life the consequences of this strange fact are intriguing but irrelevant. It may be strictly true that you are seeing your reflection in the mirror in the past, but since it takes light only one thousand millionths of a second to travel thirty centimetres (twelve inches), the delay is all but invisible. However, the further away we get from an object, the greater the delay becomes. Although over tiny distances the effect is always utterly negligible, it should be obvious that once we lift our eyes upwards to the skies and become astronomers, profound consequences await us.

RIGHT: A rare sight; in this picture Earth's crescent moon is visible above Venus (bottom) and Jupiter (right) in the night sky. As light takes longer to reach Earth from other planets and moons, depending on how far away they are, we see further into their respective pasts.

Look up at the Moon and you are looking at our closest neighbour a second in the past, because it is on average around 380,000 kilometres (236,120 miles) away; perceptible certainly, but not important. However, take a look at the Sun and you really are beginning to bathe in the past.

The Sun is 150 million kilometres away (93 million miles) – this is very close by cosmic standards, but at these distances the speed of light starts to feel rather pedestrian. We are seeing the Sun as it was eight minutes in the past. This has the strange consequence that if we were to magically remove the Sun, we would still feel its heat on our faces and still see its image shining brightly in the sky for eight minutes. And because the speed of light is actually the maximum speed at which any influence in the Universe can travel, this delay applies to gravity as well. So if the Sun magically disappeared, we would not only continue to see it for eight minutes, we would continue to orbit around it too. We are genuinely looking back in time every time we look at the Sun.

However, this is just the beginning of our time travelling. As we look up at the planets and moons in our solar system, we move further and further into the past. The light from Mars takes between four and twenty minutes to reach Earth, depending on the relative positions of Earth and Mars in their orbits around the Sun. This has a significant impact on the way we design and operate vehicles intended for driving on the surface of Mars. When Mars is at its furthest point from Earth it would take at least forty minutes to be told that a Mars Rover was driving over a cliff and then be able to tell it to stop, so Mars Rovers need to be able to make up their own minds in such situations or must do things very slowly. Jupiter, at its closest point to Earth, is around thirty-two minutes away, and by the time we journey to the outer reaches of our solar system, the light from the most distant planet, Neptune, takes around four hours to make the journey. At the very edge of the Solar System, the round-trip travel time for radio signals sent and received by Voyager 1 on its journey into interstellar space is currently thirty-one hours, fifty-two minutes and twenty-two seconds, as of September 2010.

But look beyond our solar system and the time it takes for light to travel from our nearest neighbouring stars is no longer measured in hours or days, but years. We see Alpha Centauri, the nearest star visible with the naked eye, as it was four years in the past, and as the cosmic distances mount, so the journey into the past becomes ever deeper ◉

TO THE DAWN OF TIME

When filming a series like *Wonders of the Universe*, the locations are chosen to be visually spectacular, but they must also have a narrative that enhances the explanation of the scientific ideas we want to convey. Occasionally, the locations deliver more. There is a resonance, a symbiosis between science and place that serves to amplify the facts and generates something deeper and more profound on screen. For me, the Great Rift Valley was such a place.

We arrived in Tanzania on 10 May 2010 for the first day of filming. After a brief overnight stay close to the airport at Kilimanjaro, we were driven out into the Serengeti in vintage dark green Toyota Land Cruisers, complete with exaggerated front cattle bars and shovels tied to the rear doors. The landscape is unmistakably African; the warm, damp light still wet from the rains illumines plains seemingly too vast to fit on our planet. The horizon, darkened by scattered thunderclouds stark against the early summer skies, is simply more distant than it should be. The rains have brought with them journeys, and as you drive you experience first-hand the thousand-mile migration of the Serengeti wildebeest. The relentless advance of these herds creates ruts in the drying savannah along the precise and ancient roads that always seem to run at right angles to your direction of travel, shaking the Land Cruisers to the edge of their design tolerance. Zebras, giraffe and Grant's gazelles graze, unconcerned, as our intrepid film crew rattles by.

The Great Rift Valley is not just an extraordinary geological feature ... there is more to this place because the echoes of the history of humanity ring louder across these plains than anywhere else on the planet.

LEFT AND BELOW: The Great Rift Valley, Tanzania, is one of the most spectacular geological locations on Earth. The summer skies were darkened by rainclouds, but these soon departed to reveal dusty, unmistakably African landscapes and breathtaking vistas.

Our camp is idyllic by the strictest definition of the word. Khaki tents nestle beneath acacia trees in the shadow of a giant copper-striped rock populated by a tribe of itinerant baboons intent on stealing our tape stock. Fortunately, we are guarded by the Masai, who, all cliché aside, are as tough as hell and scare not only the baboons but also the Serengeti lions and the BBC in equal amount.

So much for the visuals; the reason for the resonance of this place lies in the deep past of this dramatic landscape of life. The Great Rift Valley is not just an extraordinary geological feature that stretches 6,000 kilometres (3,700 miles) from Syria to Mozambique; there is more to this place because the echoes of the history of humanity ring louder across these plains than anywhere else on the planet. To walk this earth is to walk in the footsteps of the true ancients. Ancestors like Lucy, one of the most important fossils ever discovered, a skeleton uncovered in the Ethiopian section of the valley in 1974 by Donald Johanson. Lucy is 3.2 million years old; the remains of an *Australopithecus*, an extinct hominid species many anthropologists believe links directly to our own heritage. Further down the rift, in Tanzania, more closely related human ancestors have been discovered. In the early 1960s, Mary and Louis Leakey unearthed the remains of the earliest known species of our genus, *Homo. Homo habilis* is thought to have been a direct descendant of *Australopithecus*, and may be the first of our ancestors to have made tools. It's all in the mind, I suppose, but sitting around a fire on a cool evening in the Serengeti I felt as if I had returned to the place where I had been born after many years away. There is something about geographic origins that resonates, over a lifetime or a hundred thousand lifetimes ◉

FINDING ANDROMEDA

The connection between the history of the Serengeti and the science of light is a dimly glowing jewel in the velvet Tanzanian sky. With no cities to pollute the darkness, the plains of the African night are bathed in the light of a billion suns. The glowing arc of the Milky Way Galaxy dominates the sky, a silver mist of stars so numerous, they are impossible to count. Every single point of light and every patch of magnificent mist visible to the unaided human eye have as their origin a star in our own galaxy, or the misty clouds known as the Magellanic clouds – two small dwarf galaxies in orbit around the Milky Way. All except for one...

To find it, you first need to recognise the distinctive 'W' shape of the constellation of Cassiopeia. It sits on the opposite side of Polaris, the North Star, to the constellation Ursa Major, otherwise known as The Great Bear or The Plough. Cassiopeia, being so close to Polaris, is a constant feature in the northern skies – it simply rotates around the pole once every twenty-four hours and never sets below the horizon at high latitudes. If in your mind's eye you put the 'W' of Cassiopeia upright, then just beneath the right-most 'V' you will be able to see quite a large, faint, misty patch in the sky. It is comparable in brightness to most of the stars surrounding it, although dimmer than the bright stars of Cassiopeia. This unremarkable little patch is, in my view, the most intellectually stunning object you can see with the naked eye, because it is an entire galaxy beyond the Milky Way. It is called Andromeda, and is our nearest galactic neighbour. It is home to a trillion suns, over twice as many stars as our galaxy. It is roughly twenty-five million million million kilometres (fifteen million million million miles) away, and here is the connection.

Two and a half million years ago, when our distant relative *Homo habilis* was foraging for food across the Tanzanian savannah, a beam of light left the Andromeda Galaxy and began its journey across the Universe. As that light beam raced across space at the speed of light, generations of pre-humans and humans lived and died; whole species evolved and became extinct, until one member of that unbroken lineage, me, happened to gaze up into the sky below the constellation we call Cassiopeia and focus that beam of light onto his retina. A two-and-a-half-billion-year journey ends by creating an electrical impulse in a nerve fibre, triggering a cascade of wonder in a complex organ called the human brain that didn't exist anywhere in the Universe when the journey began ◉

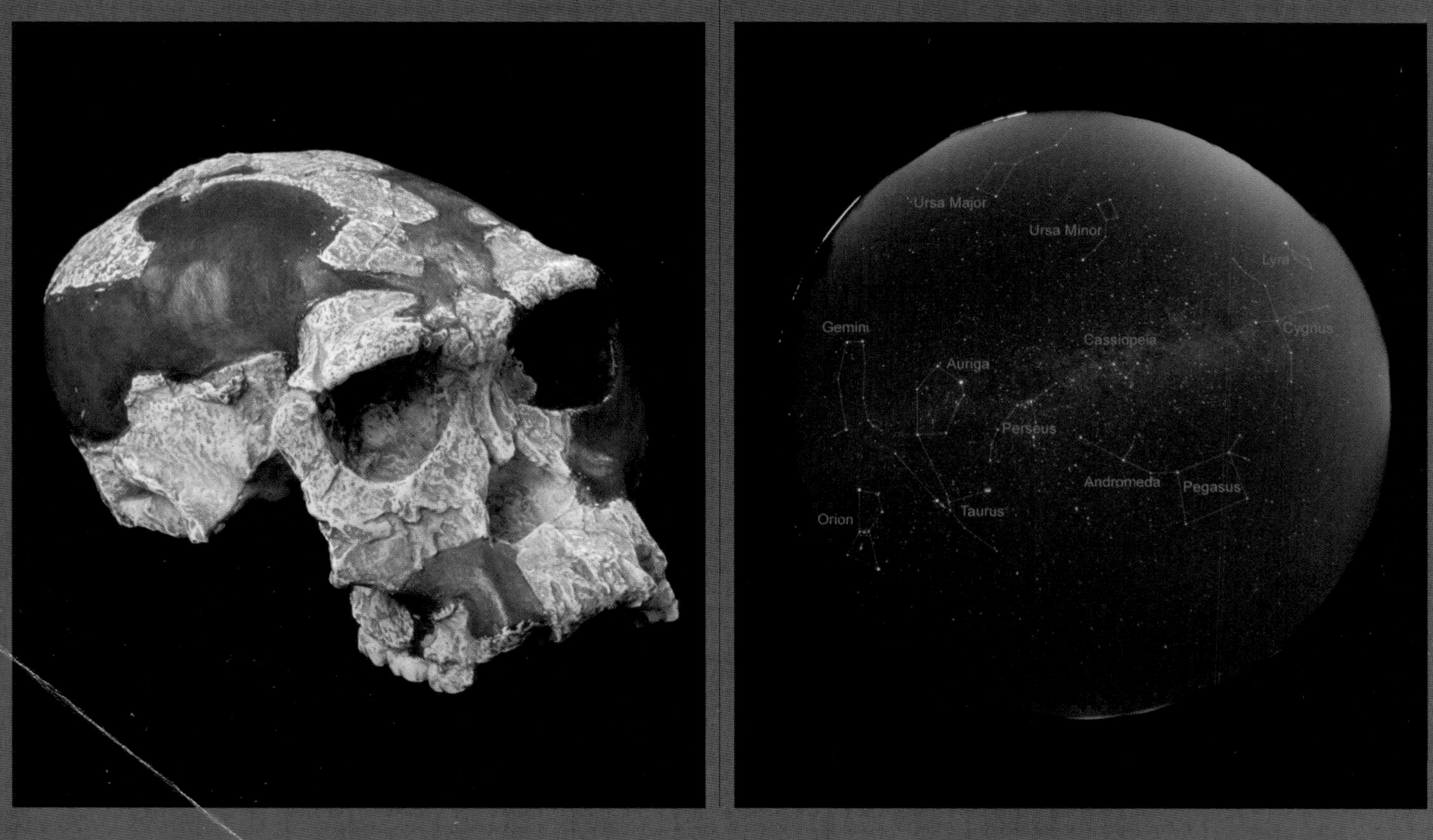

BELOW LEFT: This Homo habilis skull was found in the Olduvai Gorge in Tanzania and is believed to be around 1.8 million years old.

LEFT AND BELOW: On autumn and winter evenings, the spiral galaxy M31 (Andromeda) is visible to the naked eye in northern skies. To locate it, you first need to identify Cassiopeia, and its distinctive 'W' shape. Using the point of the 'V' on the right-hand side as an arrow, look beneath it for a large misty patch in the sky.

ALPHA CASSIOPEIAE

M31

GAMMA ANDROMEDAE

BETA ANDROMEDAE

DELTA ANDROMEDAE

ALPHA ANDROMEDAE

Observing the night skies with the naked eye can only take us so far on our journey to discover and understand the wonders of our universe. Advances in technology have brought us crafts that can take humans on expeditions beyond our planet, but also sophisticated equipment that has changed our view of the Universe entirely.

RIGHT: The Hubble Space Telescope being repaired by an astronaut from Endeavour. This eleven-tonne telescope has allowed astronomers and scientists to see further into our universe than ever before.

JPL

THE HUBBLE TELESCOPE

The naked eye can only allow us to travel back in time to the beginnings of our species; a mere 2.5 million light years away. Until recently, Andromeda was the furthest we could look back unaided, but modern, more powerful telescopes now enable us to peer deeper and deeper into space, so that we can travel way beyond Andromeda, capturing a bounty of messengers laden with information from the far distant past.

In the history of astronomy, no telescope since Galileo's original has a greater impact than the eleven-tonne machine called Hubble. The Hubble Space Telescope was conceived in the 1970s and given the go-ahead by Congress during the tenure of President Jimmy Carter, with a launch date originally set for 1983. Named after Edwin Hubble, the man who discovered that the Universe is expanding, this complex project was plagued with problems from the start. By 1986, the telescope was ready for lift off, three years later than planned, and the new launch date was set for October of that year. But when the Challenger Space Shuttle broke apart seventy-three seconds into its launch in January 1986, the shutters came down not only on Hubble, but on the whole US space programme. Locked away in a clean room for the next four years, the storage costs alone for keeping Hubble in an envelope of pure nitrogen came to $6 million dollars a month.

With the restart of the shuttle programme, the new launch date was set for 24 April 1990 and, seven years behind schedule, shuttle mission STS-31 launched Hubble into its planned orbit 600 kilometres (370 miles) above Earth. The promise of Hubble was simple: images from the depths of space unclouded by the distorting effects of Earth's atmosphere. A new eye was about to open and gaze at the pristine heavens, but within weeks it was clear that Hubble's vision was anything but 20:20. The returning images showed there was a significant optical flaw, and after preliminary investigations it slowly dawned on the Hubble team that after decades of planning and billions of dollars, the Hubble Space Telescope had been launched with a primary mirror that was minutely but disastrously misshapen. Designed to be the most perfect mirror ever constructed, Hubble's shining retina was 2.2 thousandths of a millimetre out of shape, and as a result its vision of the Universe was ruined.

Such was the value and promise of Hubble that an audacious mission was immediately conceived to fix it. This

LEFT: The Hubble Space Telescope has had a greater impact on astronomy than any other telescope. This huge telescope orbits Earth, sending back images of parts of the Universe that would otherwise remain invisible to us. The telescope has been orbiting Earth since 1990, and its revolutionary and revelatory journey continues to this day.

BELOW: The Hubble Space Telescope has brought us incredible images of other galaxies that we might never have been able to see. This shot of the spiral galaxy NGC1300 is one of the largest images taken by the telescope.

Seven years behind schedule, shuttle mission STS-31 launched Hubble ... A new eye was about to open and gaze at the pristine heavens...

was possible because Hubble was designed to be the first, and to date only, telescope to be serviceable by astronauts in space. A new mirror could not be fitted, but by precisely calculating the disruptive effect of the faulty mirror, NASA engineers realised that they could correct the problem by fitting Hubble with spectacles.

In December 1993, astronauts from the Shuttle Endeavour spent ten days refitting the telescope with new corrective equipment. In charge of the repairs, by far the most complex task ever undertaken by humans in Earth orbit, was astronaut Story Musgrave. Already a veteran of four shuttle flights, a test pilot with 16,000 flying hours in 160 aircraft types, ex-US Marine and trauma surgeon with seven graduate degrees, Musgrave is quite an extraordinary example of what people can do if they put their minds to it. He is a metaphor for the space programme itself; in Musgrave's own words, this is what restoring sight to Hubble meant. 'Majesty and magnificence of Hubble as a starship, a spaceship. To work on something so beautiful, to give it life again, to restore it to its heritage, to its conceived power. The work was worth it – significant. The passion was in the work, the passion was in the potentiality of Hubble Space Telescope.'

On 13 January 1994, NASA opened Hubble's corrected eye to the Universe and opened the eyes of our planet to the extraordinary beauty of the cosmos. A decade late and costing around $6 billion dollars, it has proved to be worth every cent ◉

HUBBLE'S MOST IMPORTANT IMAGE

LEFT: The Hubble Ultra Deep Field is one of the most spectacular and important pictures taken by the Hubble Space Telescope. This image shows nearly 10,000 galaxies of various ages, sizes, shapes and colours. The nearest galaxies appear larger and brighter, but there are also around one hundred galaxies here that appear as small red objects. These are the most remarkable features in this image; these are among the most distant objects we have ever seen.

For almost two decades the Hubble Space Telescope has captured the faintest lights and enabled us to rebuild these spectacular images, providing a window onto places billions of light years away and events that happened billions of years ago. These are places forever beyond our reach. However, there is one Hubble image that has done more than any other to reveal the scale, depth and beauty of our universe. Known as the Hubble Ultra Deep Field, this shot was taken over a period of eleven days between 24 September 2003 and 16 January 2004. During this period Hubble focused two of its cameras – the Advanced Camera for Surveys (ACS) and Near Infrared Camera and Multi-object Spectrometer (NICMOS) – on a tiny piece of sky in the southern constellation, Fornax. This area of sky is so tiny that Hubble would have needed fifty such images to photograph the surface of the Moon.

From the surface of Earth this tiny piece of sky is almost completely black; there are virtually no visible stars within it, which is why it was chosen. By using its million-second shutter speed, though, Hubble was able to capture images of unimaginably faint, distant objects in the darkness. The dimmest objects in the image were formed by a single photon of light hitting Hubble's camera sensors every minute. Almost every one of these points of light is a galaxy; each an island of hundreds of billions of stars, with over 10,000 galaxies visible. If you extend that over the entire sky, it means there are over 100 billion galaxies in the observable Universe, each containing hundreds of billions of suns.

As we stare at Hubble's masterpiece we are looking back in time; deep time, time beyond human comprehension ... the Hubble Ultra Deep Field transports us back through the history of the Universe.

However, there is something more remarkable about this image than mere scale, due to the slovenly nature of the speed of light compared to the distances between the galaxies. The thousands of galaxies captured by Hubble are all at different distances from Earth, making this image 3D in a very real sense. But the third dimension is not spatial, it is temporal. As we stare at Hubble's masterpiece we are looking back in time; deep time, time beyond human comprehension. Just as an ice core leads us back through layer after layer of Earth's history, so the Hubble Ultra Deep Field transports us back through the history of the Universe.

The photograph contains images of galaxies of various ages, sizes, shapes and colours; some are relatively close to us, some incredibly far away. The nearest galaxies, which appear larger, brighter and have more well-defined spiral and elliptical shapes, are only a billion light years away. Since they would have formed soon after the Big Bang, they are around twelve billion years old. However, it is the small, red, irregular galaxies that are the main attraction here.

There are about 100 of these galaxies in the image, and they are among the most distant objects we have ever seen. Some of these faint red blobs are well over twelve billion light years away, which means that when their light reaches us it has been travelling for almost the entire 13.75-billion-year history of the Universe. The most distant galaxy in the Deep Field, identified in October 2010, is over thirteen billion light years away – so we see it as it was 600,000 years after the beginning of the Universe itself.

It is hard to grasp these vast expanses of space and time. So, consider that the image of this ancient galaxy was created by a handful of photons of light; when they began their journey, released from hot, primordial stars, there was no Earth, no Sun, and only an embryonic and chaotic mass of young stars and dust that would one day evolve into the Milky Way. When these little particles of light had completed almost two-thirds of their journey to Hubble's cameras, a swirling cloud of interstellar dust collapsed to form our solar system. They were almost here when the first complex life on Earth arose and within a cosmic heartbeat of their final destination when the species that built the Hubble first appeared.

The story hidden within the Hubble Ultra Deep Field image is ancient and detailed, but how can we infer so much from a photograph? The answer lies in our interpretation of the colours of those distant, irregular galaxies ◉

ALL THE COLOURS OF THE RAINBOW

The breathtaking Victoria Falls are one of the most famous and beautiful natural wonders on our planet. Fuelled by the mighty Zambezi River, the falls lie on the border between Zambia and Zimbabwe in southern Africa. The falls were named by David Livingstone in 1855, the first European to see them. He later wrote: 'No one can imagine the beauty of the view from anything witnessed in England. It had never been seen before by European eyes; but scenes so lovely must have been gazed upon by angels in their flight.' That's about right from where I stood. There are few better places on Earth from which you can experience the visceral power of flowing water, but there is an ethereal feature of the falls that is just as enchanting and far more instructive for our purposes, because it holds the key to interpreting the Hubble Deep Field Image.

Hovering in the skies above the falls are magnificent rainbows, a permanent feature in the Zambian skies when the Sun shines through the mist. Rainbows are natural phenomena that have enchanted humans for thousands of years; to see one is to marvel at a simple but beautiful property of light and, as is often the case in nature, they are made more beautiful when you understand the science behind them.

Scientists have attempted to understand rainbows since the time of Aristotle, trying to explain how white light is apparently transformed into colour. Our old friend Ibn al-Haytham was one of the first to attempt to explain the physical basis of a rainbow in the tenth century. He described them as being produced by the 'light from the Sun as it is reflected by a cloud before reaching the eye'. This isn't too far from the truth. The basis of our modern understanding was delivered by Isaac Newton, who observed that white light is split into its component colours when passed through a glass prism. He correctly surmised that white light is made up of light of all colours, mixed together. The physics behind the production of a rainbow is essentially the same as that of the prism. Light from the Sun is a mixture of all colours, and water droplets in the sky act like tiny prisms, splitting up the sunlight again. But why the characteristic arc of the rainbow?

The first scientific explanation, which pre-dated Newton by several decades, was given by René Descartes in 1637. Water droplets in the air are essentially little spheres of water, so Descartes considered what happens to a single ray of light from the Sun as it enters a single water droplet. As the diagram opposite illustrates, the light ray from the Sun (S) enters the face of the droplet and is bent slightly. This is known as refraction; light gets deflected when it crosses a boundary between two different substances (point A), then when the light ray gets to the back surface of the raindrop, it is reflected back into the raindrop (point B), finally emerging out of the front again, where it gets bent a little more (point C). The light ray then travels from the raindrop to your eye (E).

The key point is that there is a maximum angle (D) through which light that enters the raindrop gets bounced back. Descartes calculated this angle for red light and found it to be forty-two degrees. For blue light, the angle is forty degrees. Colours between blue and red in the spectrum have maximum angles of reflection of between forty-two and forty degrees. No light gets bounced back with angles greater than this, and it turns out that most of the light gets reflected back at this special, maximum angle. So, here is the explanation for the rainbow. When you look up at a rainbow, imagine drawing a line between the Sun, which must be behind you, through your head and onto the ground in front of you. At an angle of forty-two degrees to this line, you'll see the so-called rainbow, or Descartes' ray of red light. At an angle of forty degrees to this line, you'll see the Descartes' ray of blue light, and all the colours of the rainbow in between. There is some light reflected back to your eye through shallower angles, which is why the sky is brighter below the arc than above it. You don't see the colours below the arc because all the rays merge to form white light. On the picture on the previous page, you can see the sky brightening inside the rainbow over the Victoria Falls, and the relative darkness of the sky outside it.

So raindrops separate the white sunlight into a rainbow because each of the consituent colours gets reflected back to your eye at a slightly different maximum angle. But why the arc? In fact, rainbows are circular. Think of the imaginary line again between the Sun, your head and the ground. There isn't just one place at which the angle between this line and the sky is forty-two degrees, there is a whole circle of points surrounding the line. The reason you can't see a complete circle is that the horizon cuts it off, so you only see the arc. This is also why you tend to see rainbows in the early morning

THE ELECTROMAGNETIC SPECTRUM
The electromagnetic spectrum is composed of a range of wavelengths from radio waves at the very longest end to gamma rays at the shortest. Our eyes are sensitive to a limited range in the middle which we know as visible light.

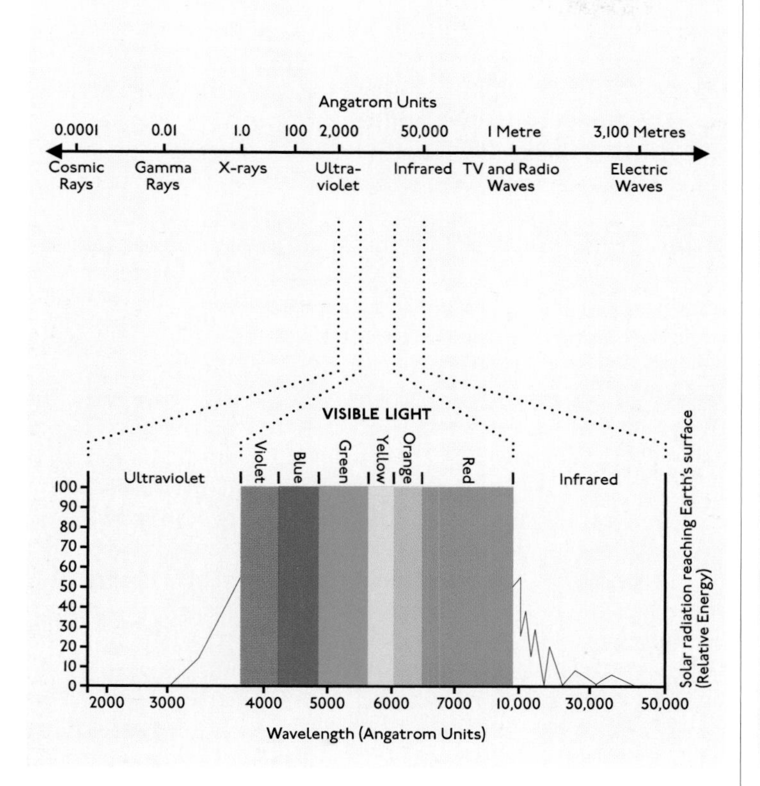

WHAT MAKES A RAINBOW AN ARC?
Decartes' theory was based on what happens to a single ray of light from the Sun as it enters a water droplet; he discovered that each colour that makes up this light is refracted, or bent, at slightly different angles to each other.

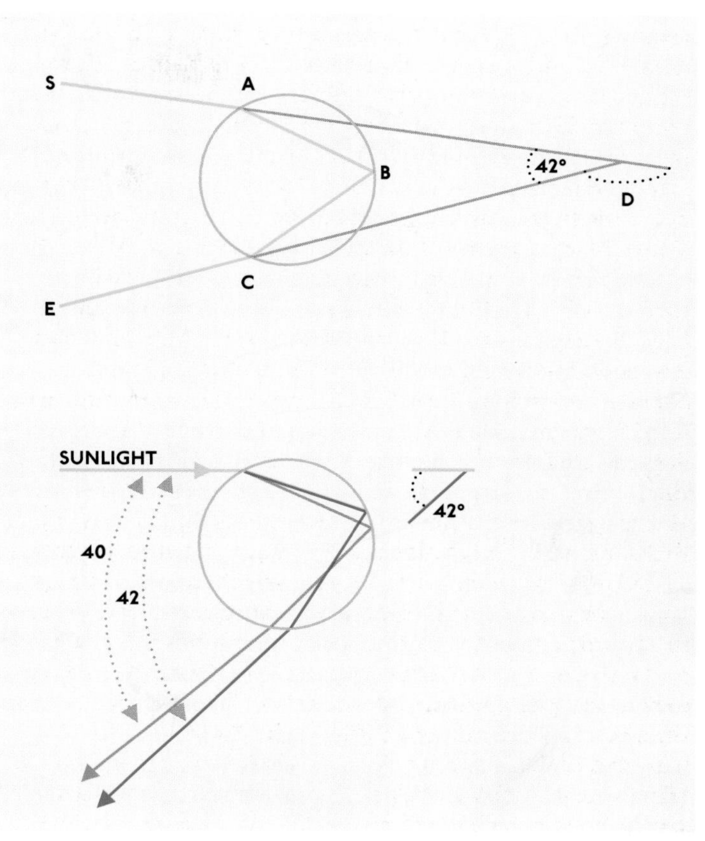

In fact, rainbows are circular. The reason you can't see a complete circle is that the horizon cuts it off, so you only see the arc. This is also why you tend to see rainbows in the early morning or late afternoon.

LEFT AND PREVIOUS SPREAD: All the way back to Aristotle, scientists have been trying to understand rainbows and how white light is transformed to colour through this medium. The Victoria Falls are perhaps one of the most spectacular places on Earth to see rainbows; here, these features hover in the sky above the cascading waters whenever the Sun shines through the mist.

or late afternoon. As the Sun climbs in the sky, the line between the Sun and your head steepens and the rainbow, which is centred on this line, drops closer and closer to the horizon until at some point it will vanish below the horizon.

These colours hidden in white light are not only revealed in rainbows; wherever sunlight strikes an object the different colours are bounced around or absorbed in different ways. The sky is blue because the blue components of sunlight are more likely to be scattered off air molecules than the other colours. As the Sun drops towards the horizon, and the sunlight has to pass through more of the atmosphere, the chance of scattering rays of yellow and red light increases, turning the evening skies redder. Leaves and grass are green because they absorb blue and red light from the Sun, which they use in photosynthesis, but reflect back the green light.

But what is the difference between the colours that makes them behave so differently? The answer goes back to our understanding of light as an electromagnetic wave. Waves have a wavelength – which is the distance between two peaks (or troughs) of the wave. Blue light has a shorter wavelength than green light, which has a shorter wavelength than red light. Our eye has evolved to discern about ten million different colours, which is to say that it can differentiate between ten million subtle variations in the wavelength of electromagnetic waves. This simple idea is all you need to read the story of the Hubble Ultra Deep Field image ◉

HUBBLE EXPANSION

BELOW LEFT: This image shows some of the most distant galaxies that we have observed – and all appear in a bright, sharp, red colour.

So, how do we know that the irregular, messy galaxies in the Hubble image are billions of light years away? The picture below shows some of the most distant galaxies we have observed. The most obvious thing about them is that they are all red. Why is this so? To answer this question correctly, we need our friend Edwin Hubble, the astronomer, again.

During the 1920s, Edwin Hubble was using what was then the world's most powerful telescope at the Mount Wilson Observatory in Pasadena, California, to observe stars called Cepheid variables. These Cepheid variables are stars whose brightness varies regularly over a period of days or months, and they are astonishingly useful to astronomers because the period of their brightening and dimming is directly related to their intrinsic brightness. In other words, it is a simple matter to work out exactly how bright a Cepheid variable star actually is just by watching it brighten and dim for a few months. If you know how bright something really is, then measure how bright it looks to you, you can work out how far away it is. Edwin Hubble's research project was simply to search for Cepheid variables in the sky and measure their distance from Earth. During his observations, he discovered two remarkable things: firstly, he quickly determined that the Cepheid variables he found in the so-called spiral nebulae (which at the time were thought to be clouds of glowing gas within the Milky Way) were in fact well outside our galaxy.

For the first time, Hubble showed that there are other galaxies in the Universe, millions of light years away.

Hubble's second observation was of even greater scientific importance. While he and others were also busy measuring the spectrum of the light from the stars in the spiral nebulae, which thanks to Hubble were now understood to be other galaxies beyond the Milky Way, they quickly observed that many of the galaxies appeared to be emitting light that was redder than it should be. Hubble quantified the amount of reddening in each galaxy as a number called redshift. Remember that red light has a longer wavelength than blue light, so seeing light redshifted simply means the wavelength is longer than expected. Hubble made his second great discovery by plotting a graph of the redshift of the light from the distant galaxies against their distance, which he had calculated from his observations of the Cepheid variables.

To his great surprise, Hubble noticed that his graph was approximately a straight line. This is because the further away a galaxy is, the greater its redshift – i.e. the more its light is stretched, and there is a very simple relationship between the distance and the redshift. Why is this? Well, the interpretation of Hubble's result is quite remarkable. The more distant the galaxy, the further the light has travelled across the Universe to reach us. Also, the further it has travelled, the more it has been stretched. This relationship

HUBBLE'S LAW: This diagram illustrates Hubble's Law; the redshift of the light from distant galaxies is plotted against their actual distance, resulting in a straight line on the graph.

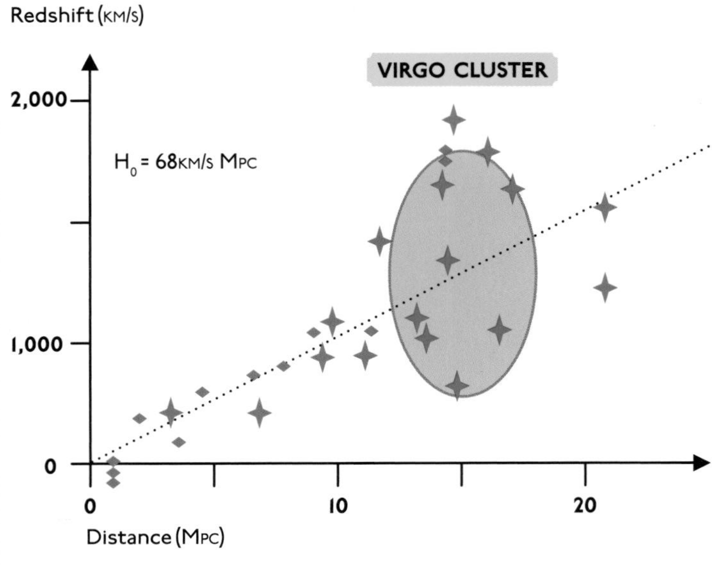

BELOW: Stephan's Quintet is a cluster of five galaxies in the constellation Pegasus, two of which, in the centre, appear to be intertwined. Studying the individual redshifts reveals that one of the galaxies is an interloper: the larger, bluer one at upper left is in fact a foreground galaxy seven times closer to us than the others. So redshifts allow us to create a three-dimensional model of the Universe.

REDSHIFT

Although first discovered in the early twentieth century, redshifts were really put into their cosmological context through the work of Edwin Hubble. He discovered that there is a very simple relationship between the distance and the redshift of a galaxy – the further away a galaxy is, the greater its redshift. This is because the further light has had to travel, the more the travelling light is stretched, and this occurs when the Universe is expanding.

REDSHIFTED SPECTRUM LINE

Other galaxy moving in expanding space

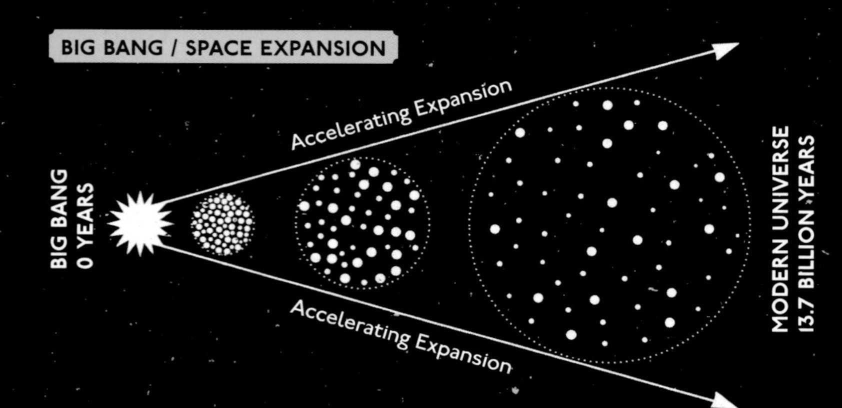

REDSHIFT

Light waves are stretched as we move away from the other galaxy in expanding space. Light is shifted to the red end of the spectrum

Our galaxy including Earth moving in expanding space

HUBBLE'S LAW

Galaxies are moving away from us with a speed that is proportional to their distance from us. The more distant the galaxy, the faster it is receding from us and the greater the redshift between distance travelled and amount of stretching occurs when something very simple but surprising is happening to the Universe. It is expanding! In other words, over the hundreds of millions of years during which the light has been travelling, space itself has been stretching at a relatively constant rate, and this has stretched the wavelength of the light in direct proportion to the distance it has had to travel. This is why the most distant galaxies have the largest redshift – their light has travelled through our expanding universe for longer and has therefore become more stretched. Hubble's discovery of this so-called 'cosmological redshift' was one of the great intellectual moments in twentieth-century science, because he discovered that we live in an expanding universe.

There is a vast amount of information contained within Hubble's simple graph. Redshift can be expressed as the amount of stretching you would see if something were flying away from you at a particular speed. The ratio of the redshift expressed in this way to the distance to the galaxy – which is the gradient of the line on Hubble's graph – is called the Hubble constant. Its value as measured today is 68 kilometres (42 miles) per second, per megaparsec. A megaparsec is a measure of distance commonly used by astronomers – 1 megaparsec is 3.3 million light years. So, another way to think of Hubble's law is that a galaxy that is 3.3 million light years away will be receding from us at a velocity of about 70 kilometres (45 miles) per second. That's pretty slow! A

BELOW: Hubble's discovery of the cosmological redshift brought about another important discovery: we are living in an expanding universe.

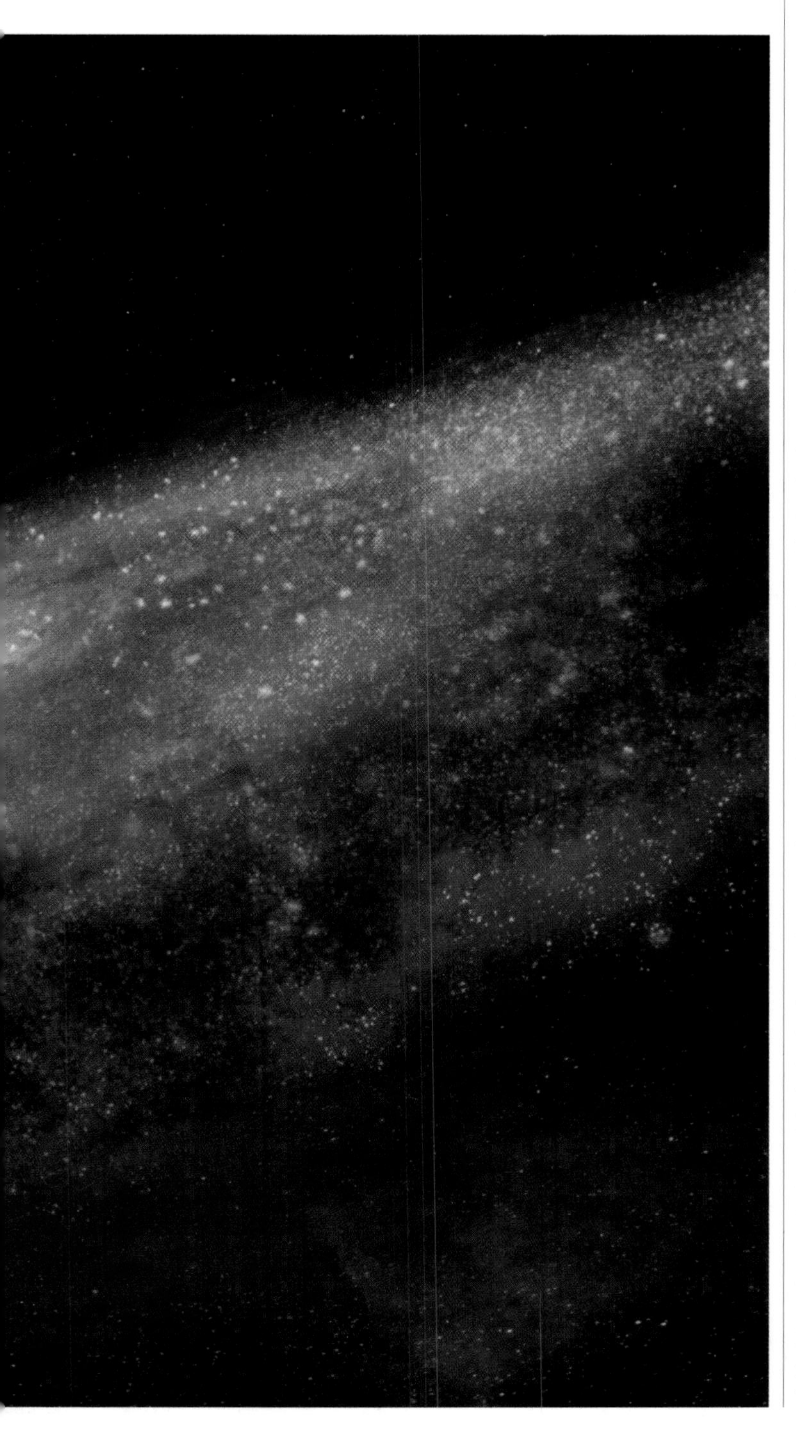

galaxy that is 6.6 million light years away will be receding at about 140 kilometres (90 miles) per second, and so on. And further, if you simply invert the Hubble constant, then you get a number with the units of time. For a Hubble constant of 70 kilometres (45 miles) per second per megaparsec, this corresponds to 14.3 billion years, which can be interpreted as the age of the Universe! (For the more mathematically inclined, you can calculate this number easily by converting megaparsecs to kilometres.) As an aside, the attentive reader might have noticed that our current best measurement for the age of the Universe is slightly lower than this, at 13.75 billion years; this is because precision measurements over the last few decades have shown us that the expansion of the Universe is not in strict accord with Hubble's simple law. The best data we have today tells us that the Universe is accelerating in its expansion due to the presence of something called dark energy.

This might seem complicated, but the conclusion is simple and profound. The reddening of the distant galaxies tells us that the Universe is expanding. This means that the galaxies we see in the sky today must have been closer together in the past. If, in your mind's eye, you keep winding back time and you watch the galaxies getting closer and closer together, then, at a time given by the inverse of the Hubble constant, you will find that they must have all been on top of each other. In other words, the Universe we see today must have been incredibly tiny. This all happened around fourteen billion years ago, and that event is what we call the Big Bang. So Hubble's remarkable observation is direct evidence that the Universe began with a big bang around fourteen billion years ago. All this was deduced in the 1920s simply by capturing the light from Cepheid variable stars and distant galaxies.

The Big Bang is difficult to visualise; it is easy to think of it as a vast explosion that flung matter out into a pre-existing void – a giant empty box, if you like – but this is completely wrong. The currently accepted picture is that all of space came into existence at the Big Bang. In fact, in the spirit of Einstein we should more correctly say that all of spacetime came into existence at the Big Bang. This means that the Big Bang didn't just happen out there somewhere in the Universe, it happened everywhere at once. So the Big Bang happened in the bit of space between you and this book; it happened inside your head, across the road, at every point in the Solar System and inside the most distant galaxies. In other words, it happened at every point in the Universe. All of space was there at the Big Bang, and all it has done is stretch ever since. This has the rather mind-bending consequence that if the Universe is infinite today, it was born infinite. Everywhere that is here now was there then, but just squashed a lot! Nobody said cosmology was easy. So when we look at the distant galaxies and we see them all flying away from us, this is not because they were flung out in some massive explosion at the beginning of time; it is because space itself is stretching, and it's been stretching since the Big Bang.

The Hubble expansion is one piece of evidence for the Big Bang, but there is another, perhaps more remarkable, fingerprint of the Universe's violent beginning, delivered to us by the most ancient light in the cosmos ●

THE BIRTH OF THE UNIVERSE

Every second, light from the beginning of time is raining down on Earth's surface in a ceaseless torrent. Only a fraction of the light present in the Universe is visible to the naked eye, though; if we could see all of it, the sky would be ablaze with this primordial light both day and night. However, some of this hidden light is not quite a featureless glow; the long wavelength universal glow known as the Cosmic Microwave Background (CMB) in fact displays minute variations in its wavelength. The CMB carries with it an image of our universe as it was just after its birth, and this discovery has provided key evidence that the beginning really did start with the Big Bang.

RIGHT: It was at the Big Bang that all of spacetime came into existence. The stars and galaxies stretched away across an infinite universe and many are still to be found today. Space is stretching still; housing the old galaxies alongside numerous new star-forming regions, such as NGC 281 k.

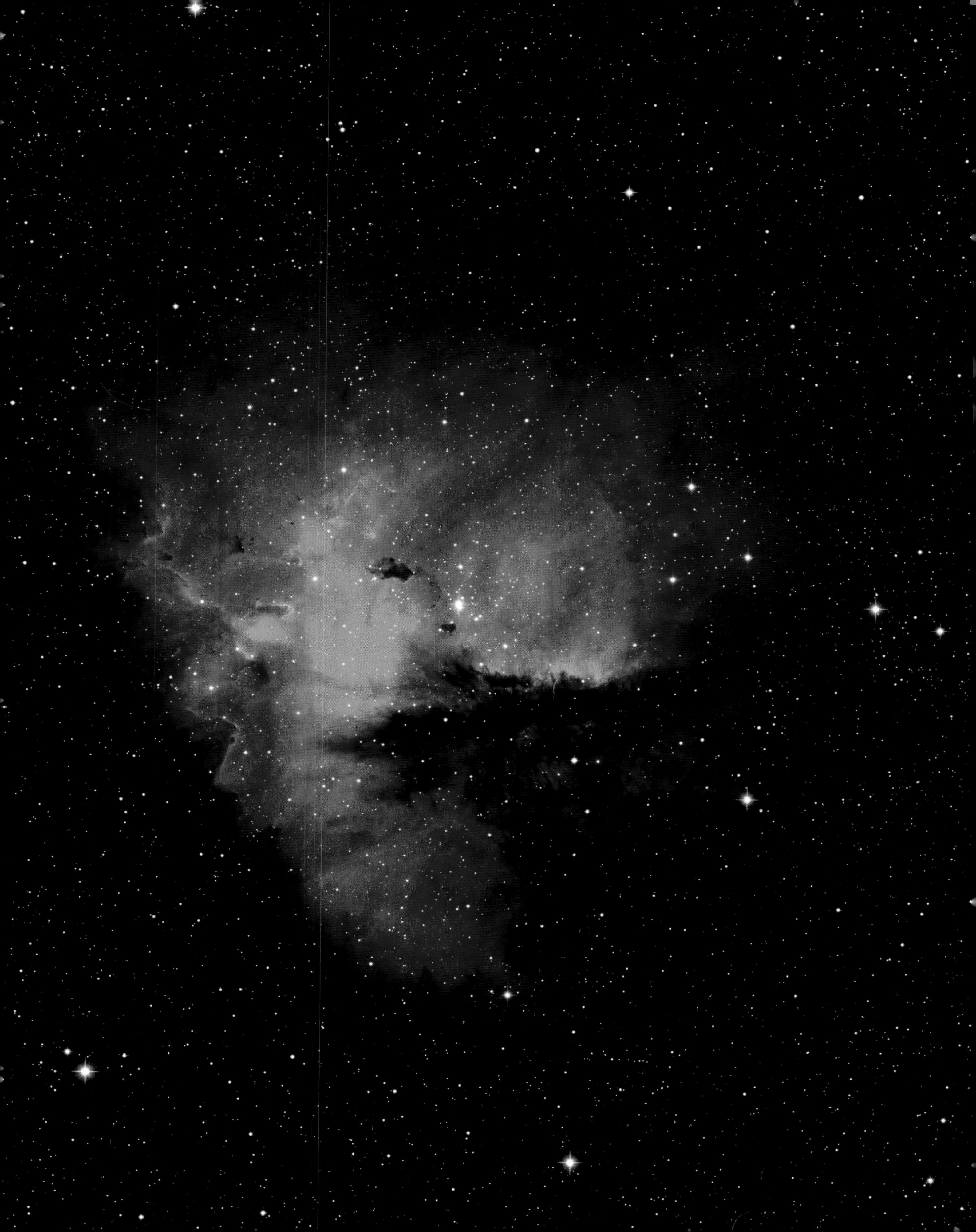

VISIBLE LIGHT

Stretching along the west coast of southern Africa is the Namib Desert. It is the oldest desert in the world; its landscape is a shifting sea of sand of over 77,700 square kilometres (30,000 square miles) which changes every minute, a consistently arid wilderness that has stubbornly avoided moisture for over fifty million years. This is a world sculpted by the Sun; its energy drives the wind that shapes the tiny grains of sand into magnificent dunes, and the colours hidden in its light paint the landscape deep orange. Yet even when the Sun has set, the desert remains awash with light and colour, but the human eye can't see it.

Visible light is a tiny fraction of the light in the Universe. Beyond the red, the electromagnetic spectrum extends to wavelengths too long for our eyes to detect. It's still light; still the sloshing back and forth of the electric and magnetic fields driving forwards through the void at the special universal speed, it's just we didn't evolve to see it. In the Namib Desert you can feel this light, though, if you hold your palms towards the sand. The dunes are warm long after sunset, and this residual heat is nothing more than long-wavelength light. A scientist would call it infrared light; the only difference between infrared and visible light is the wavelength – infrared has a longer wavelength than visible light. Travel further along the spectrum, past infrared, and we arrive at microwaves, with wavelengths unsurprisingly about the size of a microwave oven. The spectrum then seamlessly slides into the radio region, with wavelengths the size of mountains.

Throughout most of human history we have been blind to these more unfamiliar forms of light, but to detect them you don't need expensive, hi-tech kit, just a radio. When tuning a radio you are not tuning into a sound wave, you are picking up information encoded in a wave of light. Most of the radio waves we are familiar with are artificially created and used for communication and broadcasting, but just as there is plenty of visible light in the Universe that isn't manmade, so there are naturally occurring microwaves and radio waves too. And just like the visible photons from the most distant galaxies, the microwave and radio photons are messengers, carrying detailed information about distant places and times across the Universe and into our technologically created artificial eyes.

Next time you are tuning a radio and can hear static, you are actually listening to a deeply profound sound – you are listening to the Big Bang.

Next time you tune a radio, listen to the static between the stations. About 1 per cent of this is music to the ears of a physicist because it is stretched light that has travelled from the beginning of time. Deep in the static is the echo of the Big Bang. These radio waves were once visible light, but light that originated 400,000 years after the Big Bang. Prior to

LEFT: Standing among the dunes of the Namib Desert you become aware of the sheer scale of the landscape. It is a landscape sculpted by the Sun and coloured by it at all times.

BELOW: Forget state-of-the-art kit, all you need to use to detect hidden forms of light is a simple radio. As you tune, it you will pick up information encoded in a wave of light.

BOTTOM: Only a fraction of light is visible in the Universe. This infrared image shows the massive scale of the Universe and demonstrates how the electromagnetic spectrum extends to wavelengths that are too long for our eyes to detect. Here we can see hundreds of thousands of stars at the core of the Milky Way Galaxy, but so many are still hidden from our view.

that, the observable universe was far smaller and hotter than it is today. At 273 million degrees Celsius, this is an order of magnitude hotter than the centre of a star, so hot that the hydrogen and helium nuclei then present in the Universe couldn't hold onto their electrons to form atoms. The Universe was a super-heated ball of naked atomic nuclei and electrons known as a plasma. Light cannot travel far in dense plasma because it bounces off the electrically charged subatomic particles. It was only when the Universe had expanded and cooled down enough for the electrons to combine with the hydrogen and helium nuclei to form atoms that light was free to roam. This point in the evolution of the Universe, known as recombination, occurred around 400,000 years after the Big Bang, when the Universe had cooled to about 3,000 degrees Celsius and was around a thousandth of its present size. That is close to the surface temperature of red giant stars, so the whole Universe would have been glowing with visible light like a vast star. The Universe has become cooler and more diffuse since, so this ancient light has been free to fly through space, and it is some of these wandering messengers that we collect with a detuned radio today. However, as the Universe has expanded, space has stretched and so too has the light – so much so that the light is no longer in the visible part of the spectrum. It has moved beyond even the infrared, and is now visible to us only in the microwave and radio parts of the spectrum. This faint, long, wavelength universal glow is known as the Cosmic Microwave Background, or CMB, and its discovery in 1964 by Arno Penzias and Robert Wilson was key evidence in proving that the Universe began in a Big Bang ◉

PICTURING THE PAST

BELOW: This detailed picture of the Universe in its infancy was pieced together from data collected over several years by the Wilkinson Microwave Anisotropy Probe (WMAP). The different colours reveal the 13.7-billion-year-old temperature fluctuations that correspond to the seeds from which the galaxies grew.

On 30 June 2001, the Wilkinson Microwave Anisotropy Probe, known as WMAP, was launched from the Kennedy Space Center in Florida. This highly specialised telescope was built with a single purpose: to capture the faint glow of the CMB and create the earliest possible photograph of the Universe. After nine years of service, WMAP has recently been retired, but its photograph is still the object of frenzied research because it contains so much rich detail about the early Universe and its expansion and evolution ever since.

This much-studied image is probably the most important picture of the sky ever taken. It may not look like much; it doesn't have the beauty of a spiral galaxy or nebula, but to a scientist it is the most beautiful picture ever taken because it contains a vast amount of information about the history of our Universe.

The raw image from WMAP shows the glow of our Milky Way Galaxy as it creates a hot bright band across the sky, but once this detail and other observational side-effects are removed, we are left with this simplified, but equally important and informative, picture below. This photograph of the night sky documents in extraordinary detail the structure of our universe at the time of recombination. Over the nine years in which WMAP was in service, the detail of this image has been repeatedly refined, which in turn reveals more and more detailed information encoded in the primordial light.

The WMAP data is presented as a temperature map of the sky. The wavelength of the detected light at any particular point corresponds to a temperature; shorter wavelengths are higher temperatures, longer wavelengths are lower ones. The red areas are hotter than the blue, but only by around 0.0002 degrees. The average temperature of the CMB is 2.725 degrees above absolute zero. On the Kelvin temperature scale, that's 2.725 K, or -270.425 Celsius.

Despite being incredibly tiny, these temperature differences are of overwhelming importance because they tell us that in the very first moments of our universe's life there were regions of space that were slightly denser than others. These virtually imperceptible differences might not seem much, but without them we would not exist. That's because these little blips in the CMB are the seeds of the galaxies. The red spots in the CMB correspond to parts of the Universe that were on average around half a per cent denser than the surrounding areas at the time of recombination. As the whole Universe expanded, these areas would have expanded slightly more slowly than their surroundings because of their higher density – effectively, their increased gravity due to their higher density would have slowed the expansion, causing their density to increase further relative to the space around them. By the time the Universe was one-fifth of its present size, just over a billion years after the Big Bang, these regions would have

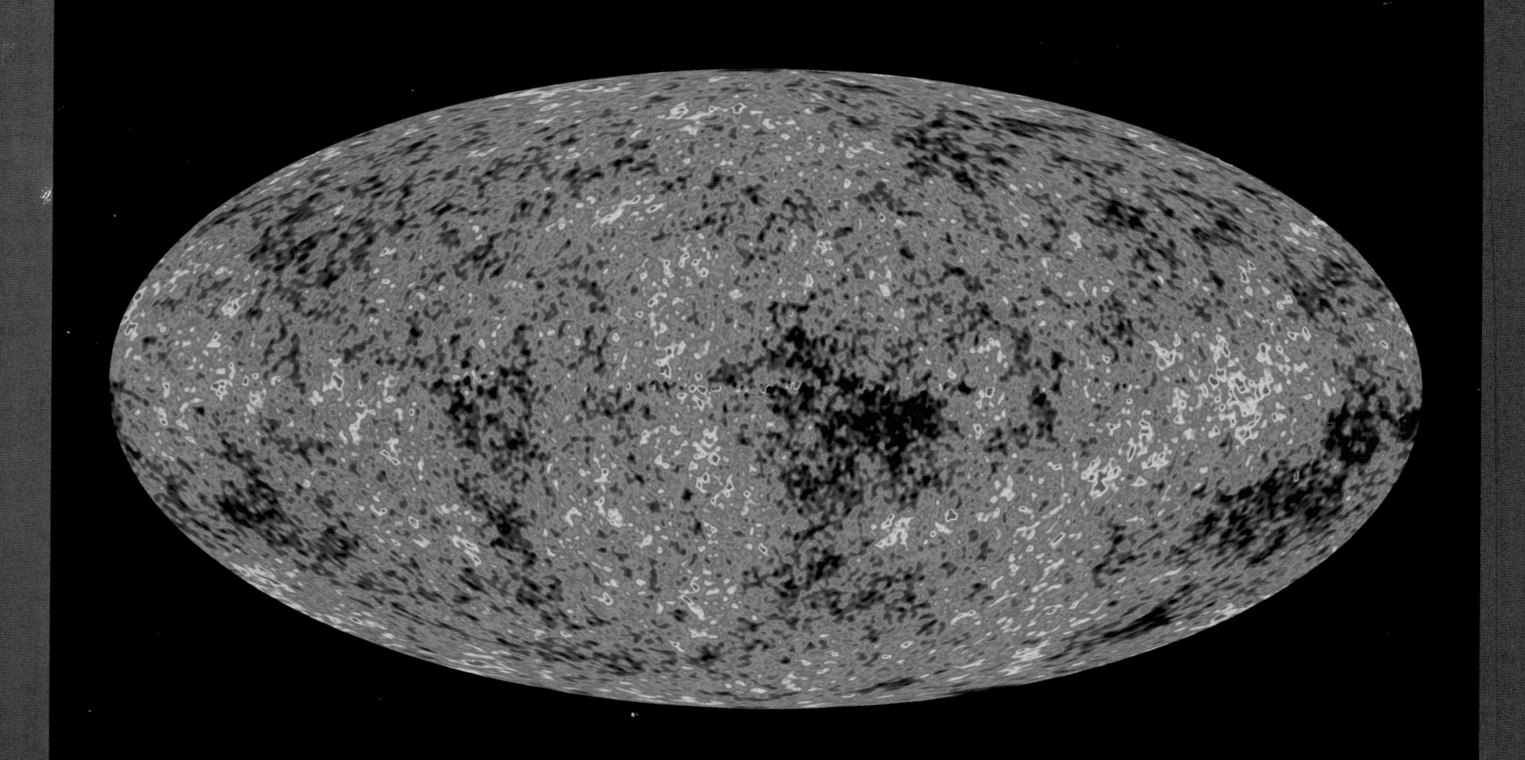

BELOW: As the Universe expanded, the denser areas within it expanded more slowly than others because of their increased gravity. By the time these areas were twice as dense as their surroundings, the matter within them was sufficiently cool and dense to collapse under their own gravity and form the first stars and cores of new galaxies.

been twice as dense as their surroundings. By this time the matter in these regions was dense enough and cool enough to begin to collapse under its own gravity, leading to the first star formation and the emergence of the cores of the galaxies, including our own Milky Way. This is the cosmic epoch we see in the most redshifted Hubble Space Telescope data – the formation of the first galaxies – and their seeds are the minute fluctuations visible in the Cosmic Microwave Background Radiation.

The rest, as they say, is history. Across the cosmos, countless suns began to switch on and to fill the Universe with light. For billions of years, generations of stars lived and died until, 9 billion years after it all began, in an unremarkable piece of space known as the Orion Spur off the Perseus Arm of a galaxy called the Milky Way, a star was born that became known as the Sun. This is the story of how our solar system has its ultimate origin in those dense areas of space that appeared in the first moments of our Universe's life. But what is the origin of those tiny fluctuations in density that we see in the CMB?

This is perhaps the most remarkable piece of physics of all. The most popular current model for the very very early Universe is known as inflation. The idea is that around 10–36 seconds after the Big Bang, the Universe went through an astonishingly rapid phase of expansion in which it increased in volume by a factor of around 10^{78}! In less scientific notation, that's a million million million million million millionths of a second after the Big Bang, and an increase in volume by a factor of a million billion billion billion billion billion billion billion billion billion. This was all over by 10^{-32} seconds or so. Before inflation, the part of the Universe we now observe, all the hundreds of billions of galaxies in our night sky, would have been far, far smaller than a single subatomic particle. At these minute distance scales, quantum mechanics reigns supreme, and tiny quantum fluctuations before inflation would have been magnified by the rapid expansion to form the denser regions we observe in the Cosmic Microwave Background spectrum. If inflationary theory is correct, the CMB is therefore a window onto a time in the life of the Universe far earlier than 400,000 years after the Big Bang. We are seeing the imprint of events that happened in the truly mind-blowing first million million million million million millionths of a second after it all began. I find this the most astonishing idea in all of science. From a vantage point of 13.7 billion years, little beings like you and me scurrying around on the surface of a rock on the edge of one of the galaxies are able to understand the evolution of the Universe and speculate intelligently about the very beginning of time itself, just by decoding the messages carried to us across the cosmos on beams of light. The power of science is quite genuinely daunting, the richness of its stories unparalleled, the cosmos it reveals, beautiful beyond imagination.

There is one last twist to this story. Throughout our journey, light has been the messenger, carrying stories of far-flung places and the distant past to our shores. But there is evidence from one of the ancient sites on our home planet that light may have played a far more active role in our history than mere muse ◉

FIRST SIGHT

Hidden in the high Rocky Mountains in British Columbia, Canada, is one of the most important and evocative scientific sites on Earth, and it's where the story of light and our lives begins. Around 505 million years ago, when this whole area lay deep beneath the surface of a primordial ocean, it was hit by a huge mudflow. The mud buried everything in its path and created a snapshot of a remarkable time in the evolution of life on Earth. A whole ancient ecosystem was frozen and preserved intact in the mud; the lives of the primitive creatures documented by a chance geological event with the care and precision with which the Egyptians created their glorious tombs half a billion years later. For hundreds of millions of years, this ancient treasure trove was locked away, but in 1909 it was uncovered high on a mountainside. This is the Burgess Shale.

The Burgess Shale is one of the most important fossil sites in the world. It is not just the number and diversity of the animals found here, it's their immense age. Before around 540 million years ago, there are no fossils of complex life forms found anywhere on the surface of Earth. We know that there was life before this period, but the animals were very simple creatures that didn't possess skeletons of any kind. This means that they don't show up on the fossil record. In the geological blink of an eye in the period of time immortalised in the Burgess Shale, known as the Cambrian Era, it appears

One current theory for the origin of the Evolutionary Big Bang is that the emergence of the eye in animals such as the trilobite triggered the Cambrian Explosion. Once one predatory species develops eyes, there is a powerful selection mechanism in favour of others developing and refining eyes too.

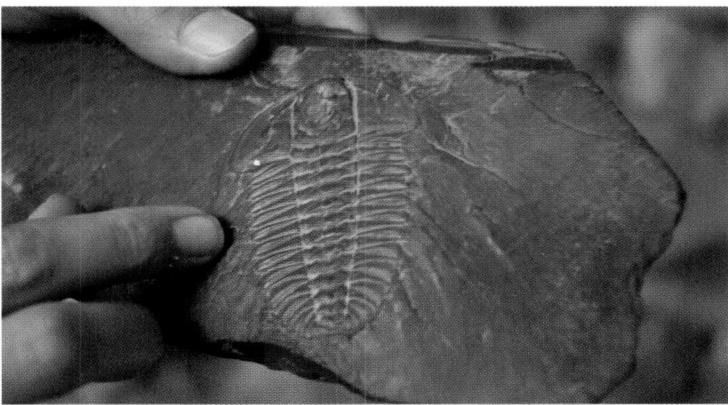

LEFT AND TOP: The Burgess Shale is one of the most important and exciting fossil sites in the world, where a staggering amount of diverse animals are to be found, dating back over 500 million years.

ABOVE: Numerous genera of trilobites have been found in the Burgess Shale. These fossils are so detailed and well preserved that they have enabled scientists to make important observations about the structure and behaviour of these now-extinct organisms.

that a vast range of complex multi-cellular life emerged on the planet. Biologists call it the Evolutionary Big Bang, or the Cambrian Explosion.

So what triggered the evolution of complex life? There is a clue in these fossil beds that lie high in the Canadian Rocky Mountains. The picture left shows one of the ancient animals found here; a complex organism called a trilobite. Trilobites, now long extinct, had external skeletons and jointed limbs, but most strikingly they had complex, compound eyes. These prehistoric predators could see shapes, detect movement and use their eyes very effectively to chase their prey. The ability to see made these trilobites very successful animals indeed; in fact they survived for a quarter of a billion years, only vanishing from Earth in the Permian mass extinction 250 million years ago.

One current theory for the origin of the Evolutionary Big Bang is that the emergence of the eye in animals such as the trilobite triggered the Cambrian Explosion. Once a predator possesses eyes which will help it chase its prey, a new force in natural selection is immediately introduced. The animals that survive this selection are those that are best adapted to this new threat; they may camouflage themselves, leading to an increasingly sophisticated visual appearance, or dodge the predators with enhanced sense organs. In other words, once one predatory species develops eyes, there is a powerful selection mechanism in favour of others developing and refining eyes too. In turn, this selects far more sophisticated predators, and so on. This is in a sense an evolutionary arms

race, as the pressure of natural selection leads more and more complex life forms to develop.

These early creatures, immortalised in the Burgess Shale, were among the very first to harness the light that filled the Universe. Before they emerged, the rise and fall of the Sun and the stars in the night sky went unnoticed. These creatures are our ancestors, and in fact there is also evidence at Burgess that we humans may only exist because of one particular adaptation in a strange, worm-like creature called a Pikaia. Although the Pikaia looks unimpressive, it may be one of the most important animals ever discovered. It is thought by some, although not all, evolutionary biologists that the Pikaia is the earliest known ancestor of modern vertebrates – the branch of life that we are categorised in – so it could be that this little worm-like creature is our earliest known ancestor. What is also fascinating about Pikaia is that it may have had light-sensitive cells that allowed it to evade predators and survive in the Cambrian seas – cells that may have evolved over many hundreds of millions of years into our eyes. This is all speculative, but it is possible that without Pikaia's primitive yet remarkable ability to detect the light from the Sun, we humans may never have appeared on planet Earth. Perhaps there would never have been a life form

We have even been able to capture the light from the beginning of time and we have glimpsed within it the seeds of our own origins.

here with the ability to do the one thing that has allowed us to understand our universe more than anything else: to look up.

Understanding the Universe is like reading a detective story, and the essential evidence we need to solve it has been carried to us across the vast expanses of space and time by light. We have even been able to capture the light from the beginning of time and we have glimpsed within it the seeds of our own origins. We've seen things our ancestors wouldn't believe: stars being born in distant realms, and galaxies lost in time at the very edge of the visible Universe and our cosmos just moments after it all began.

It's a wonderful thought that these primitive biological light detectors that emerged on Earth half a billion years ago in the Cambrian Explosion have evolved into those most human of things; our green, blue and brown eyes that are able to gaze up into the night sky, capture the light from distant stars and tell the story of the Universe ◉

LEFT: The Carina Nebula is a large bright nebula that surrounds several clusters of stars. It contains two of the most massive and luminous stars in our Milky Way galaxy, Eta Carinae and HD 93129A. Located 7500 light years away, the nebula itself spans some 260 light years across, about seven times the size of the Orion Nebula, and is shown in all its glory in this mosaic. It is based on images collected with the 1.5-metre Danish telescope at ESO's La Silla Observatory.

CHAPTER 2

STARDUST

THE ORIGINS OF BEING

What are we made of? This is an old question, maybe one of the oldest, and one that thinkers and scientists have been working hard to answer since ancient times. This work continues today, and it may be that by the time you read this book the story of the search for the building blocks of the Universe will have another chapter. Such is the power, excitement and rate of progress of modern science. This chapter is the story of how those building blocks were created in the very early Universe, fused into more complex structures over billions of years in the furnaces of space, and delicately assembled by the forces of nature into planets, mountains, rivers and human beings.

RIGHT: The Large Hadron Collider (LHC) is the highest energy particle accelerator at CERN (the European particle physics laboratory near Geneva, Switzerland). In this huge machine, 27km (17 miles)in circumference, proton beams are accelerated so that they collide head-on. The resultant particles can be detected and recorded so that scientists can then try to understand how they fit together.

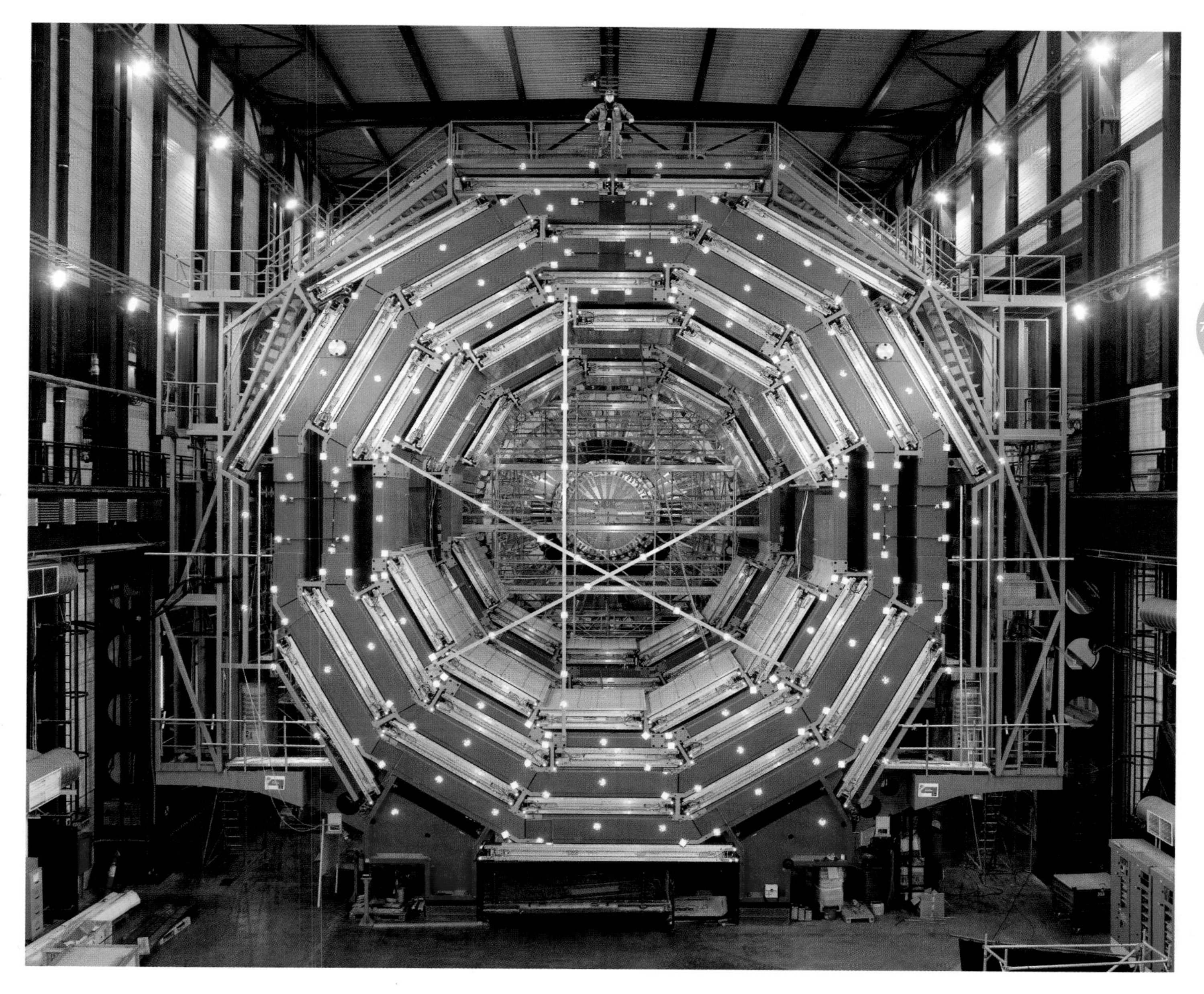

The ancient Greeks thought deeply about the question of what we are made of, although they lacked the scientific methodology and technology to arrive at a definitive answer. This led to many competing hypotheses, including some that got close to our modern view: we are all made out of smaller pieces. That there are the smallest building blocks of matter (indivisible basic units that can be fitted together to build the world) was termed the 'atomic hypothesis', a theory usually credited to two thinkers – Leucippus and Democritus – in around 400 BC. They held that the world was created from an infinite number of different types of indivisible and indestructible atoms. Each had a different shape, allowing them to fit together neatly to build large objects. So, iron was made of one type of atom, water of another, human flesh of another, and so on. They thought atoms possessed the properties of their real-world substances – water atoms were slippery, while metal ones were shaped so that they locked together to produce very hard substances. We now know that this is not only wrong, but a gross overcomplication. While their hypothesis correctly stated that the world is made from smaller pieces, you don't need an infinite number of atom types to build the complexity around us. A human is made of the same stuff as a rock; a fish of the same stuff as the Earth; the sky of the same stuff as the oceans. Enumerating the basic building blocks and understanding how they fit together is the province of the science of particle physics, and this quest continues at the Large Hadron Collider at CERN, in Geneva.

By early 2011, we had discovered that the Universe is composed of twelve basic building blocks, only three of which are required to build everything on our planet, including our bodies. These three components, known as the up and down quarks and the electron, can be assembled into the more familiar protons and neutrons – two up quarks and a down quark make a proton, and two down quarks and an up make up a neutron. In turn, the protons, neutrons and electrons make up the chemical elements – ninety-four of which are known to occur naturally – including the basic chemical elements hydrogen, carbon, oxygen, iron, gold and silver ◉

THE CYCLE OF LIFE

BELOW: The Bagmati River is lined with funeral pyres burning the bodies of the deceased.

Fifteen miles northeast of the Nepalese capital city of Kathmandu, three small streams come together to mark the beginning of one of the holiest rivers in the world. At its source the Bagmati is a fast-running mountain stream, but by the time it winds through the Kathmandu valley and enters the great city of the Himalayas it has become a wide and majestic river.

In the eastern part of the city, where the river's mythical power is at its greatest, stands the fifth-century Pashupatinath Temple, one of the most sacred sites in the Hindu world. Pilgrims come from all over India and Nepal to worship there and pay their respects to the god Shiva.

I have always found the Hindu faith fascinating; it is rich and complex, a disorientating mix of mythology and philosophy, a continual and jagged juxtaposition of temples, holy sites, rituals and everyday life that produces a joyful assault on the senses. Pashupatinath is no exception. It is at once vibrant and ethereal, a place where the colours and noise of India meet the gentle philosophy of Tibet and the hybrid dissolves into the crystal-clear, high Himalayan air in the smoke of a thousand burning bodies on the funeral pyres lit at this holy place. The scent of burning flesh mixes with incense and tinkling bells, and the sound of chanting Monkey Gods continually interrupts the calls of market traders.

A central tenet of Hindu philosophy is the concept of the Trimurti – the triad of the three fundamental aspects of the Supreme Being, represented as the great gods Brahma, Vishnu and Shiva. Lord Brahma is the creator of the Universe, Lord Vishnu the preserver, and Lord Shiva the destroyer. Shiva represents darkness, as an angry god who will eventually bring an end to Earth, yet in Hinduism this destruction is seen as an essential part of the cycle of life, because in order for new things to be created, the old order must first be destroyed. Shiva is therefore also a regenerative

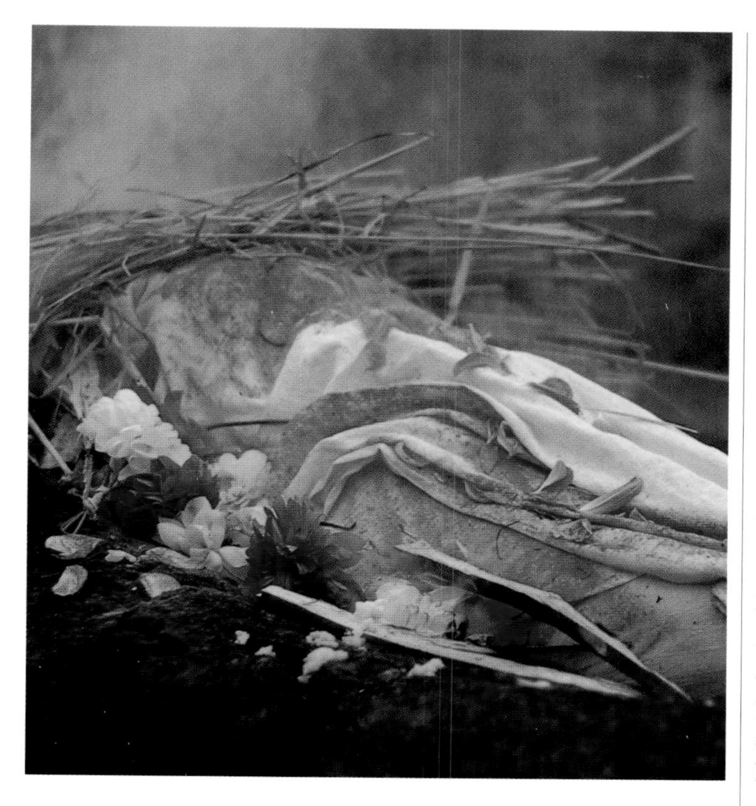

LEFT: For Hindus, the passing of a loved one is a stage in the endless cycle of death and rebirth that is central to their beliefs. Cremations are a familiar sight along the holy Bagmati River; the body is dipped in the river three times before cremation, and at the end of the ceremony the chief mourner must bathe in the river's water, often accompanied by the other attendant mourners.

or reproductive power, part of the endless cycle of death and rebirth that is central to the Hindu belief system. This is why the Pashupatinath Temple and the river it stands beside are revered as places to die.

Hindus believe that the purpose of a soul's time on Earth is to work through a cycle of rebirth and reincarnation until it becomes perfect. Only then can it be reunited with the Universal Soul and be freed from its material existence. The *Bhagavad Gita* says: 'Just as a man discards worn-out clothes and puts on new clothes, the soul discards worn-out bodies and wears new ones'. By having your body cremated on the riverbank beside Shiva's Pashupatinath Temple, it is believed that your soul will be released from the worn-out body as quickly and easily as possible.

According to the Nepalese Hindu tradition, the dead body must be dipped three times into the Bagmati River before cremation. The chief mourner, usually the first son of the deceased, lights the funeral pyre and must bathe in the waters of the holy river immediately after the cremation. Many of the relatives who join the funeral procession also bathe in the river or sprinkle the holy water on their bodies. This makes the river bank a strange and crowded place. To my British eyes it is somewhat shocking, because death is rarely, if ever, paraded like this; but here in Kathmandu it is not seen to be insensitive to wander between the pyres as the relatives and friends go through their rituals.

In Hindu tradition the human body consists of five elements: air, water, fire, earth and ether. Remarkably, according to modern science, this is overcomplicated, but their belief about what happens to these elements after death parallels our modern understanding of how the world works.

Underlying the cremation ceremony is the conviction that the elements of the body vacated by the soul are returned to Earth to be re-used and recycled. Death is therefore not an end for the immortal soul or the mortal flesh, it is simply the conclusion of one stage of existence and the beginning of another; part of a natural cycle of death and rebirth. As far as the atoms and molecules in our bodies are concerned, modern science is in complete agreement with that idea. When I die my constituents aren't going to be magically destroyed; they will be returned to Earth and, given enough time, they will become part of some other structure.

Of course, Hinduism isn't alone in having rich and lyrical stories about the origin and evolution of Man and the Universe. Virtually every society and every religion around the world has at its heart a creation story that explains where we come from, how we came to be here, and what will happen to us when we die. This suggests that curiosity about our origins is an innate, perhaps even a defining part, of the human condition.

Underlying the cremation ceremony is the conviction that the elements of the body vacated by the soul are returned to Earth to be re-used and recycled . . . As far as the atoms and molecules in our bodies are concerned, modern science is in complete agreement with that idea.

In common with the great systems of thought throughout history, modern science has its own creation story to tell – one based on physics and cosmology. It can tell us what we're made of and where we came from – in fact, it can tell us what everything in the world is made of and where it came from. It also answers that most basic of human needs: to feel part of something much bigger, because to tell this story you have to understand the history of the Universe. It also teaches us that the path to enlightenment is not in understanding our own lives and deaths, but in understanding the lives and deaths of the stars ◉

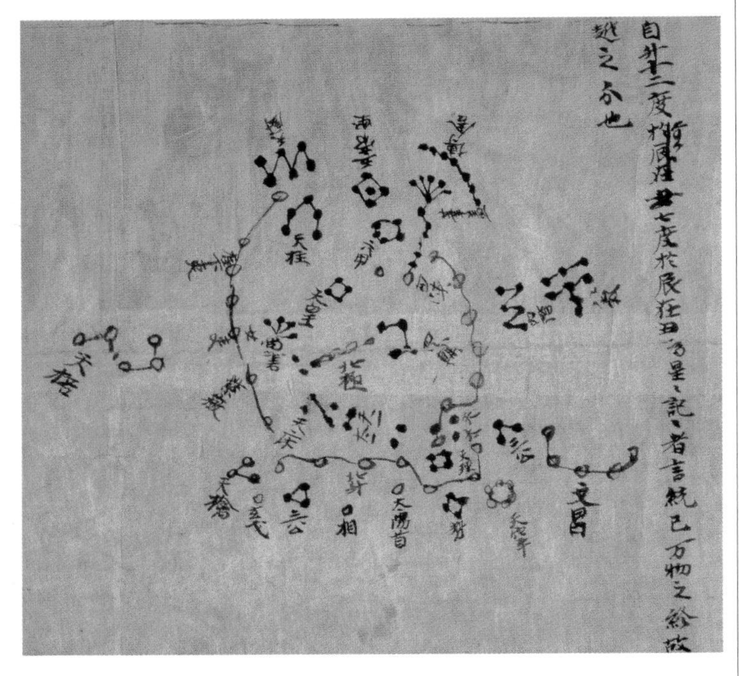

BELOW: The Dunhuang star chart dates back to AD 700 and is the oldest existing star chart. It was named after the place where it was found along the Silk Road trade route in northern China (in the twentieth century) and is now owned by the British Library. It depicts the stars in the sky according to the Chinese constellation tradition.

RIGHT: This celestial map shows a more detailed, highly illustrated view of the constellations according to Dutch cartographer Frederik de Wit in the seventeenth century.

MAPPING THE NIGHT SKY

The moment you leave a city and experience a truly dark night sky, it becomes obvious why our ancestors spent a great deal of time looking up at the stars. They are a bewildering array; a patterned silver canopy self-evidently not devoid of meaning or purpose. For thousands of years ancient astronomers endeavoured to capture and catalogue every light; to observe, log and name as many of these distant suns (for we now know their true nature) as they could. The oldest-known record of a star chart may be over thirty thousand years old. A carved ivory mammoth's tusk, discovered in Germany in the late 1970s, appears to be imprinted with a pattern that resembles the constellation of stars we now call Orion. In France, cave paintings have been discovered which reveal that humans were mapping the night skies tens of

For thousands of years ancient astronomers endeavoured to capture and catalogue every light; to observe, log and name as many of these distant suns (for we now know their true nature) as they could.

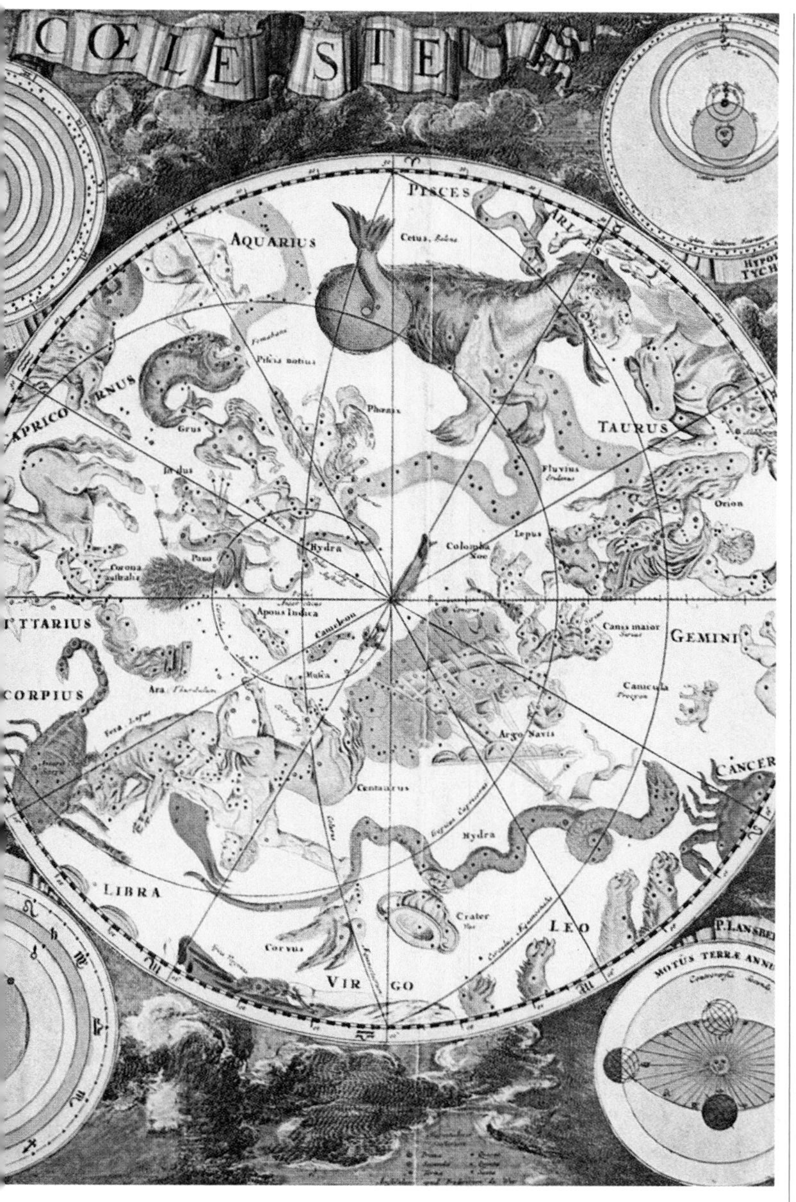

BELOW: In AD 185, Chinese astronomers witnessed a brightness in the sky comparable to that of Mars, and this remained for eight months. This phenomena was the first recorded occurrence of a supernova explosion, but it was not until late 2006 that the remains of this cosmic event were identified. This picture, taken by the Chandra X-ray Observatory, shows an object now known as RCW 86. The image shows low-, medium- and high-energy X-rays in red, green and blue respectively. It was the study of the distribution of X-rays with energy, combined with measuring the remnant's size, that enabled scientists to conclude that RCW 86 was created by the explosion of a massive star around 8,000 light years away.

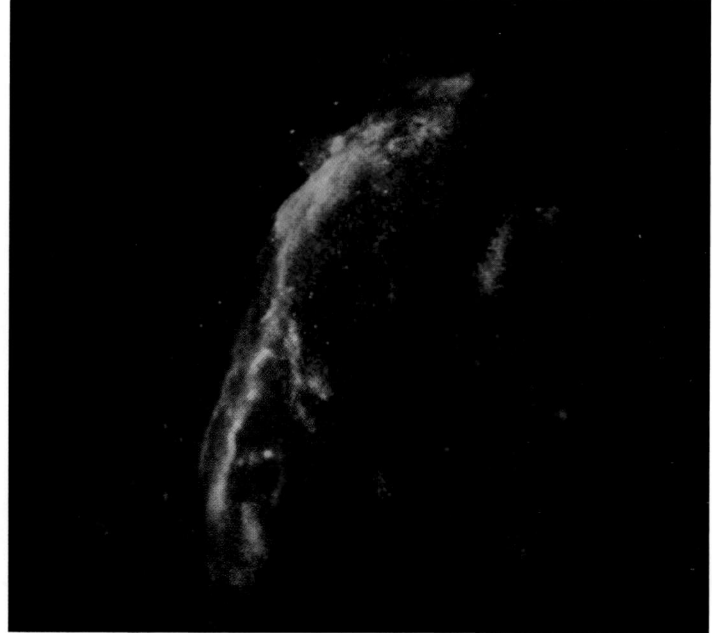

thousands of years before the great civilisations of antiquity began to slowly explore the Universe in more detail.

The Egyptians were one of the first ancient cultures to not only map the night sky but to name some of the stars they observed. They called the North Star the 'star that cannot perish', and they also recorded the names of constellations. The Sumerians and Babylonians went a step further by writing down these early names and patterns and creating astronomical catalogues that listed and grouped stars in ever-increasing complexity. Greek, Chinese and Islamic astronomers all continued to build ever more complex systems of classification, with many stars today still being referred to by their original Arabic names.

To the ancients, the stellar backdrop had a deceptive permanence that no doubt motivated them to record and mythologise the patterns they saw. But in AD 185, for the first time in recorded history, a particular type of fleeting addition to the lights in the night sky was observed and documented. Understanding the nature of this rare and spectacular phenomenon eventually led us beyond merely naming the stars and enabled us to tell the story of their births and deaths.

In late 2006, the remains of the cosmic event of AD 185 that illuminated the skies and minds of Chinese astronomers almost two thousand years ago was identified. The picture above, taken by the Chandra X-ray Observatory, is that of the object known as RCW 86. This object is thought to be the still-glowing remains of one of the most powerful events in our universe – a supernova explosion.

Supernovae are the final act in the lives of massive stars, colossal explosions in which a single star can shine as brightly as a billion suns. If RCW 86 is the remains of the AD 185 supernova, then the 'guest star' described by the Chinese astronomers that glowed brightly in the skies for eight months before fading from view was around 8,000 light years away – a quite colossal distance for something to shine so brightly in our skies. The ancient astronomers didn't know it at the time, of course, but they had documented the first clear evidence that the stars must all eventually die ◉

STELLAR NURSERIES

BELOW: These fascinating ultraviolet images, taken by NASA's Galaxy Evolution Explorer, show a star named Mira speeding across the sky and leaving behind it an enormous trail of debris. This material is in fact 'seeds' which will be recycled to create new stars, planets and possibly even life, as it travels through our galaxy.

Above our heads a story of life and death is being told in spectacular fashion. This tale begins in the vast stellar nurseries where new stars burst into life. These fertile areas of star formation are known as nebulae and are among the most beautiful structures in the skies. One of these, the Orion Nebula (pictured far right), is perhaps the most studied astronomical object. It is usually credited as being discovered by Nicolas-Claude Fabri de Peiresc in 1610, but there is evidence from folk tales that the Mayans knew of the faint smudge beneath the stars of Orion's belt. It can be seen with the naked eye in a very dark clear sky, and it is this complex, ever-changing formation that has taught us most about how stars are born.

The Omega Nebula (the Horseshoe, or Swan, Nebula) is a vast interstellar cloud that is over fifteen light years across and illuminated by hundreds of bright young stars. These stars, depending on their masses, will burn for hundreds of millions or billions of years, sending a constant stream of light across the Universe until their voracious hunger depletes the hydrogen in their cores and forces them to expand and transform into giants.

As they near the end of their lives, the most massive stars are transformed into colossal giants – such as the red Mira, whose radius is 400 times that of our sun and only just clinging onto life. When the end finally comes for stars like these, the ensuing supernova explosion will leave only a faint trace of the star. For the largest stars, the supernova will leave a black hole behind – an object so dense that even light cannot escape its clutches. Slightly smaller stars will end their post-supernova days as neutron stars, which we detect by the lighthouse beam of radiowaves they emit as they spin every few seconds or less.

Stars much smaller than Mira won't go out with a bang. Such relatively cool stars are called red dwarfs and are the most common type of star in our galaxy. Perhaps the most famous of these we have studied is Gliese 581. Just over twenty light years away from Earth, this star has been the subject of intense observation in recent years due to the discovery of at least six exoplanets orbiting around it. Most excitingly, planet Gliese 581 g is thought to orbit within the habitable zone of the star and so is considered a prime location for the search for extraterrestrial life

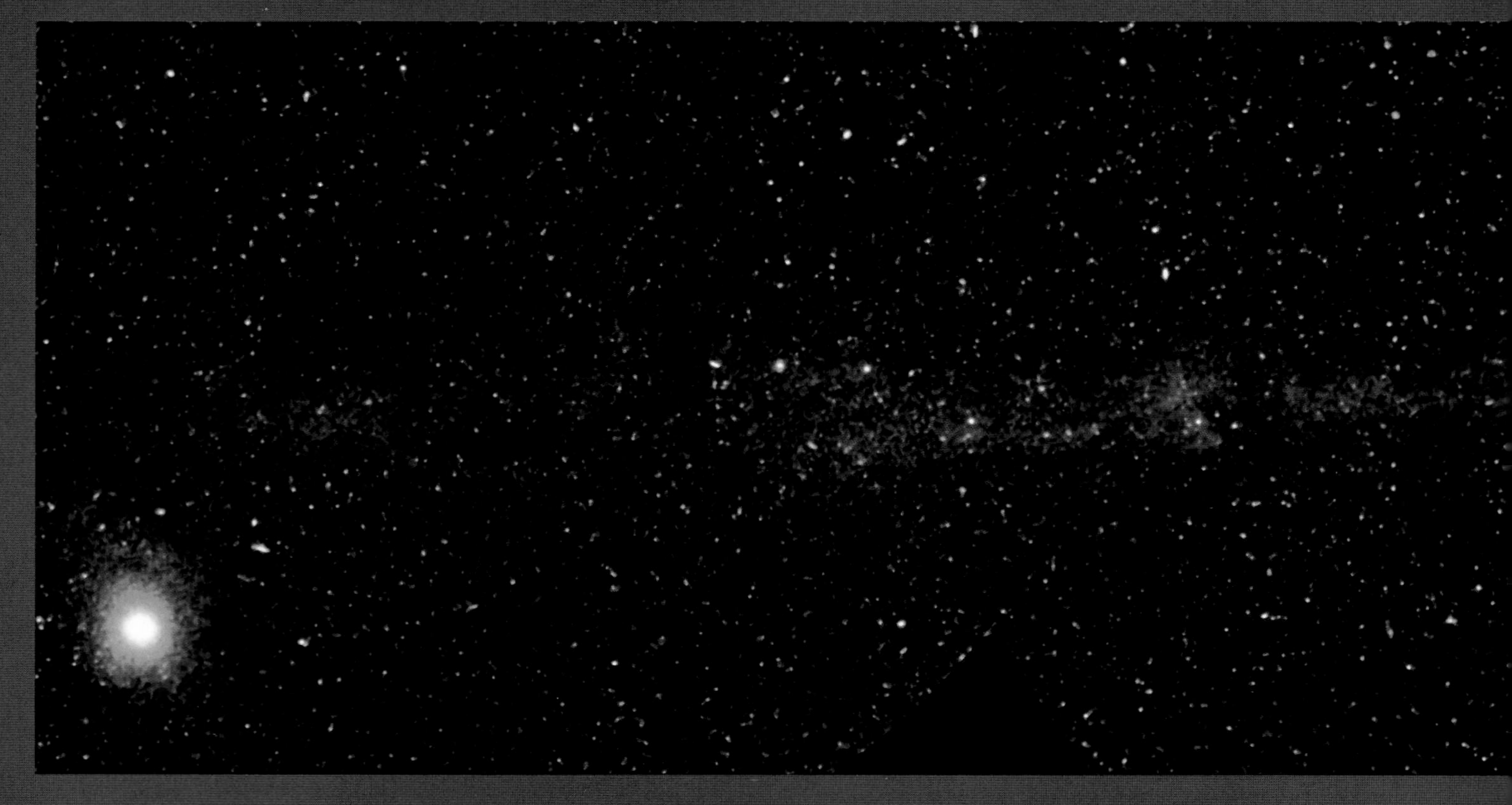

BELOW: Known as the Horseshoe, or Swan, Nebula, this molecular cloud is also often called the Omega Nebula, due to its similarity in shape to the Greek letter Omega. Ultraviolet light from a cluster of massive young stars buried within the nebula make the surrounding gas glow. This image was taken by the European Southern Observatory's 3.6-metre (11.8-foot) telescope in La Silla, Chile.

BELOW: Perhaps the most studied astronomical object, the Orion Nebula is also one of the most beautiful structures in the sky. On over 100 orbits of Earth between October 2004 and April 2005, NASA's Hubble Space Telecope captured this nebula in one of the most detailed astronomical images ever produced. On a clear, dark night sky this impressive formation – which includes more than 3,000 stars of varying sizes – can be seen with the naked eye. This complex, constantly evolving formation has provided scientists with crucial insight into how stars are formed.

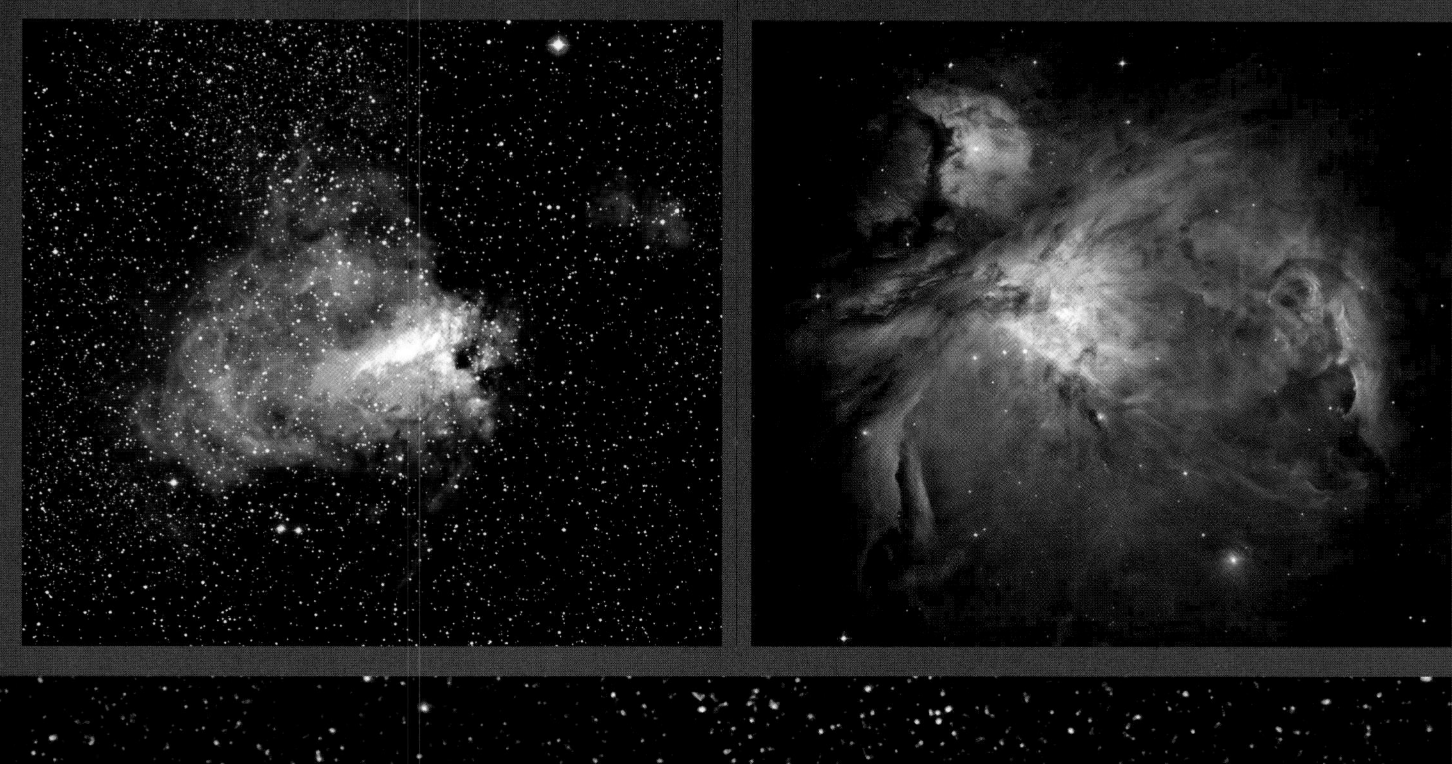

The Venus transits of our sun are a rare occurrence, they only happen twice in eight years and won't be repeated for another 100 years. The last transit happened in 2004, with another due in 2012. We will have to wait until 2116 for the next one.

RIGHT: This is a composite image of Venus transiting the Sun on 8 June 2004. Venus can be seen from Earth as a small black disc moving across the face of the Sun.

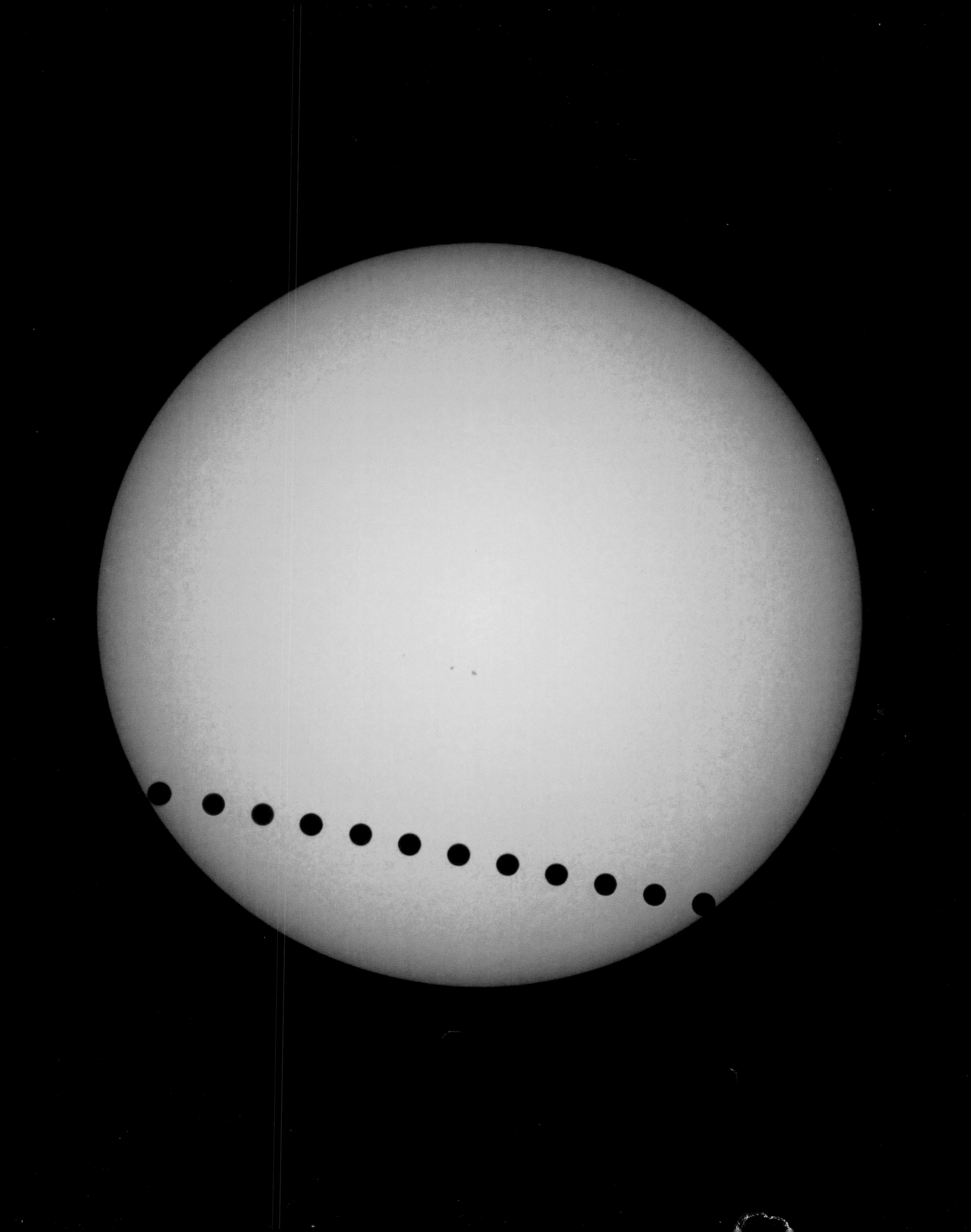

HOW TO FIND EXOPLANETS

One of the most exciting areas of current astronomical research is the hunt for planets around other stars – known simply as exoplanets – which are potential homes for extraterrestrial life. Until recently such a search would have been impossible, as planets are too faint to see over interstellar distances, however, thanks to new modern instrumentation, we are now able to detect the

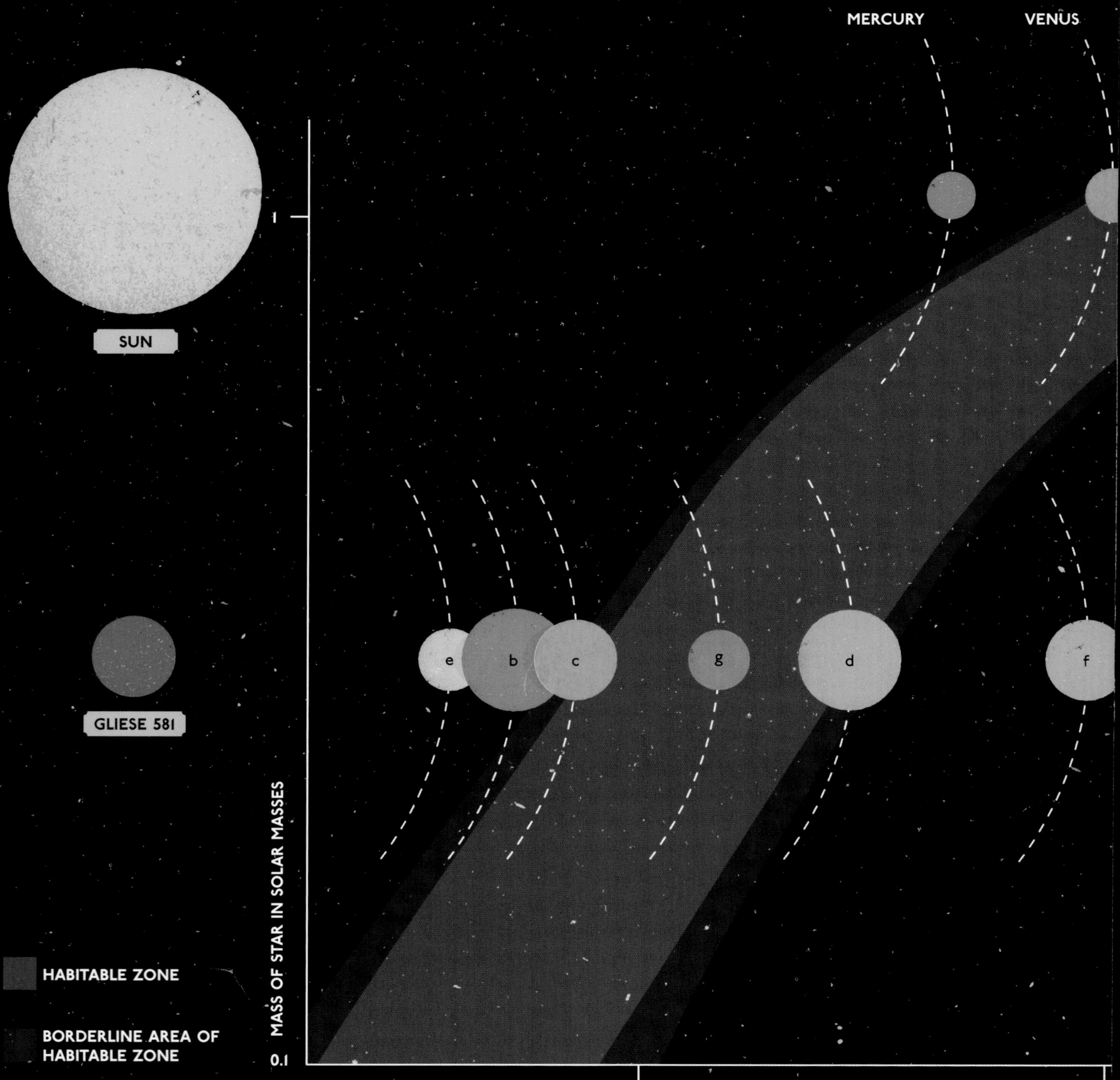

SUN

GLIESE 581

MERCURY

VENUS

e b c g d f

1

0.1

MASS OF STAR IN SOLAR MASSES

HABITABLE ZONE

BORDERLINE AREA OF HABITABLE ZONE

telltale signals of exoplanets using two main techniques: the radial velocity method and the transit method. With these techniques, individual planets and even planetary systems have been discovered around hundreds of stars. Masses of these extrasolar planets range from a few times that of Earth, to the size of 25 Jupiters. Whether a planet could support life depends on its distance from the parent star. Around each star is a 'habitable zone', in which temperatures are suitable for water to exist as a liquid. The size of this zone depends on the energy output of the star; the faintest ones have the closest, narrowest zones. The red dwarf Gliese 581 is believed to have at least one planet within its habitable zone.

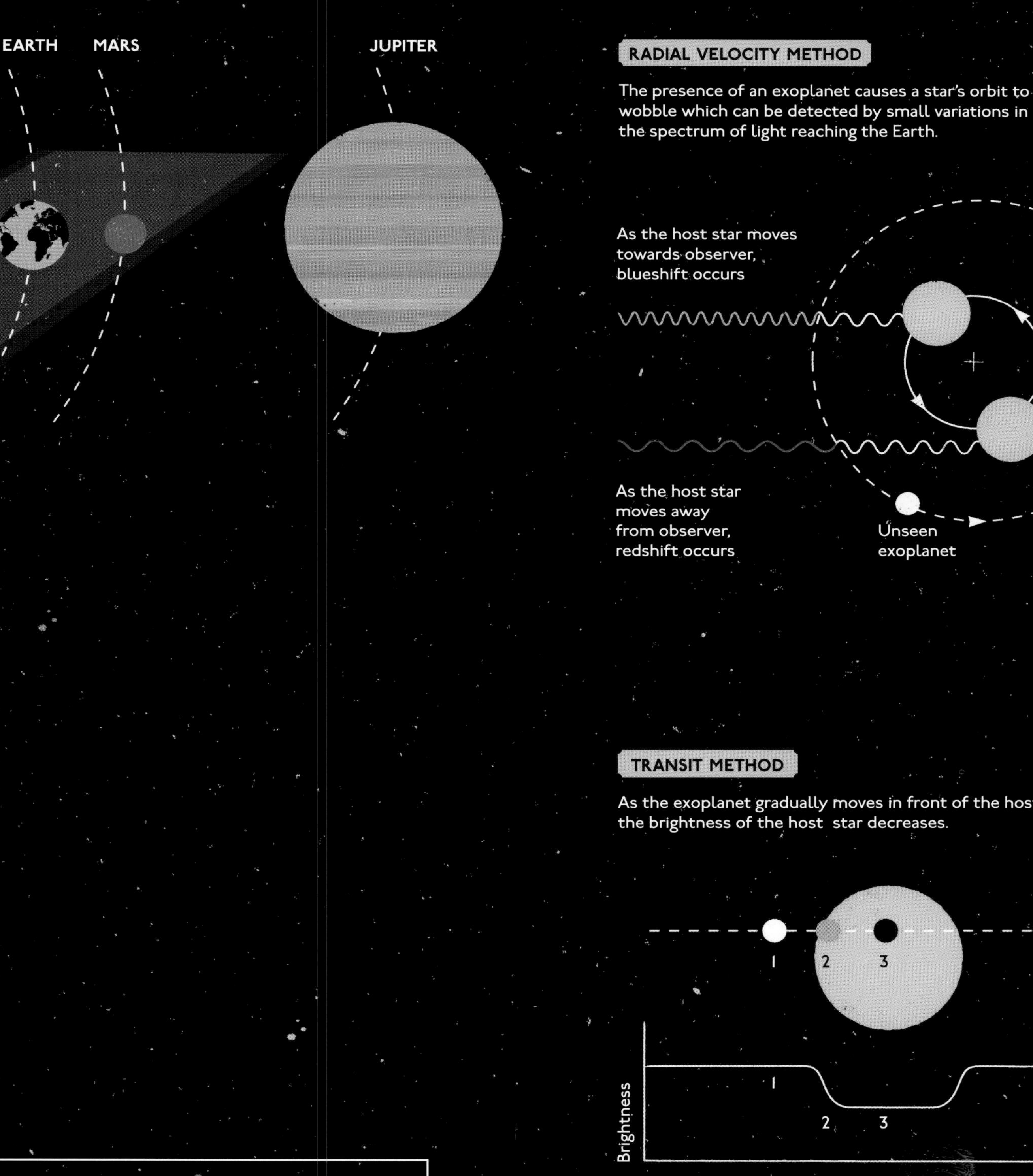

RADIAL VELOCITY METHOD

The presence of an exoplanet causes a star's orbit to wobble which can be detected by small variations in the spectrum of light reaching the Earth.

TRANSIT METHOD

As the exoplanet gradually moves in front of the host star, the brightness of the host star decreases.

As we observe all the cosmic structures around us in spectacular detail, each tells us something different about the life cycle of the stars. However, something much deeper can be learnt from understanding the existence of stars: they are the ultimate origin of all but the simplest of Leucippus' and Democritus' long-sought-after atoms, and as such are the building blocks of ourselves. To comprehend how the stars could play such a vital role in our existence, we must momentarily step back from the skies and come firmly back down to Earth.

THE ORIGINS OF LIFE

The first step in understanding how the lives of stars are precursors to our own lives is to discover exactly what we are made of. There is possibly no more beautiful, and perhaps no more instructive, place on Earth to begin this journey than in the shadow of the world's tallest mountain range. With over 100 peaks exceeding 7,200 metres (23,620 feet), the Himalayan range is truly a land of giants; nine of the ten highest mountains on Earth are part of the Himalayas. The greater Himalaya is home to forty-five of the world's top fifty highest peaks. Spectacularly beautiful, it is the sheer scale of these mountains that hides a fascinating and instructive first step on the road to understanding the building blocks of the Universe. Despite their majesty, just a few tens of millions of years ago these mountains were something very different.

As well as being the largest mountain range on the planet, the Himalayas is also one of the youngest. Just seventy million years ago (a very short time in geological terms) the Himalayas didn't exist. The relentless movement of Earth's tectonic plates shaped these mountains in a geological heartbeat. As the Indo-Australian plate collided with the Eurasian plate at the rate of about 15 centimetres (6 inches) a year, the ocean floor in between began to crumple and rise up to form the mountain range. This means that much of the rock out of which these towering peaks are made was formed at the bottom of an ocean, only to be lifted up thousands of metres into the air over a few short millions of years.

The evidence for this extraordinary journey is not difficult to find. If you look closely at any piece of Himalayan limestone you will see it has a chalky, granular structure. What you are looking at are the petrified remains of sea creatures – the bodies and shells of coral and polyps that died millions of years ago in a long-lost ocean. Given a relatively short timescale and a bit of pressure, these biological remains are quickly converted into solid rock. Limestone can also be formed by the direct precipitation of calcium carbonate from water, although the biological sedimentary form is more abundant. We know that the Himalayan limestone is predominantly biological because we have found fossils at the top of Mount Everest! There is perhaps no better example of the endless recycling of Earth's resources that has been going on since its formation almost five billion years ago.

We humans are also very much part of that system. As unsettling as it may sound, every atom in your body was once part of something else. It may have made up an ancient tree or a dinosaur, and you'll be pleased to know it was certainly part of a rock. The reason this can happen – that the rocks of Earth can become living things and that living things will eventually die and become rocks again – is simple: everything in the Universe is composed of the same basic ingredients ◉

LEFT: The largest mountain range in the world, the Himalayas is also the youngest. This panorama, taken from the top of Kala Pattar in the Sagarmatha National Park, Nepal, shows only a fraction of its scale. Understanding the creation of these impressive mountains helps us to answer many questions about the structure of all living elements in the Universe.

BELOW LEFT: When you are presented with the sheer magnitude of the Himalayas and the towering peak of Mount Everest, it is hard to believe that these huge mountains started off life at the bottom of an ocean.

BELOW RIGHT AND BOTTOM: Natural recycling at its most impressive. The Himalayan limestone has been proved to be predominantly biological, due to the quantity of fossils of sea shells and creatures that have been found at the summit of Mount Everest.

IMAGE COURTESY OF PERIODICTABLE.COM

THE PERIODIC TABLE

For many people the Periodic Table will provide a strong echo of the school science laboratory. At its simplest, this chart is a list of the chemical elements, fundamental units of matter, which were considered to be the smallest building blocks of the world. However, this table is much more than just a list. Although elemental theories of matter were first postulated in Greece, it wasn't until 6 March 1869 that the Russian chemist Dmitri Mendeleev finally tamed the ever-expanding list of the basic constituents of matter. Mendeleev's genius was to arrange the list of the sixty-six then-known elements into a table according to their chemical properties. In the process, the table not only provided a neat way of grouping the elements according to their properties, but also predicted the existence of eight elements yet to be discovered. Over the next thirty years, all eight were discovered, including gallium and germanium, and were found to have the exact properties predicted by Mendeleev's table. The number of elements continued to grow, and by 1955 the one-hundred-and-first element was discovered (named Mendelevium as a tribute to the father

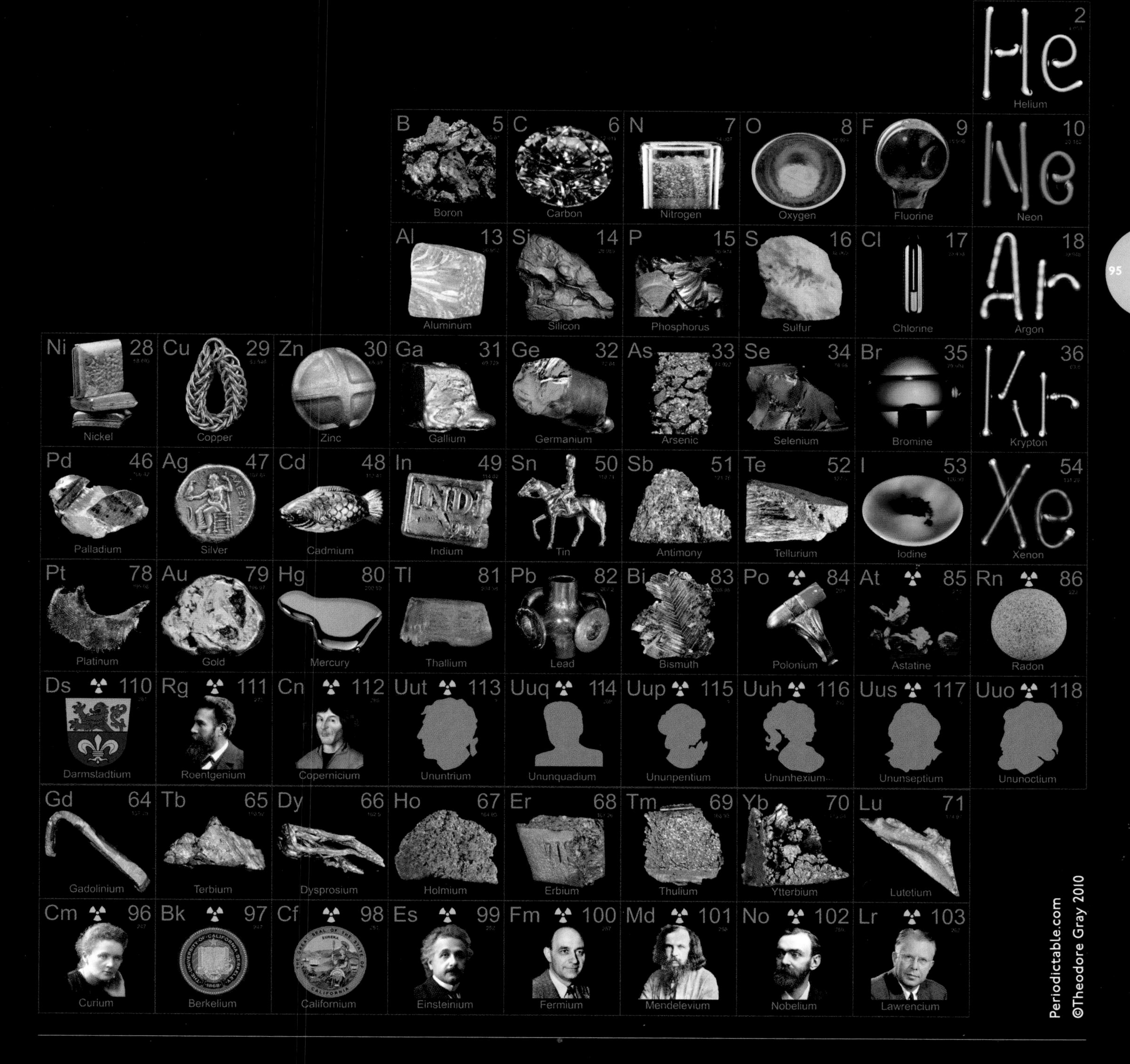

of the Periodic Table) by a group of scientists at the University of California, Berkeley. To date, 118 elements have been categorised, the latest of which, ununseptium, was successfully synthesized and detected by a Russian–US team in April 2010.

Starting with hydrogen and ending with plutonium, the first ninety-four elements of the Table have been found occurring naturally on Earth. These elements are nature's building blocks; the remaining twenty-four elements, can only be created artificially and live for very short periods of time. Using these ninety-four elements you can explain all of biology and chemistry without knowing about the underlying structure of protons and neutrons, electrons and quarks. This is because you need very high energies and temperatures to break apart the elements – a condition that only exists naturally deep inside the stars.

The first step of our journey to explain where we come from is to understand the origin of these ninety-four elements. But first we must discover how we know that everything we see in the sky is made of the same stuff as us on the ground ◉

THE UNIVERSAL CHEMISTRY SET

Surprising as it sounds, we know what every star, planet and moon in the observable Universe is made of, despite the fact that there is only one other place in the Universe that humans have actually visited in person.

On 21 July 1969, Neil Armstrong and Buzz Aldrin became the first humans to set foot on another world. They spent 2 hours, 36 minutes and 40 seconds walking on the surface of the Moon, but it wasn't until the last half hour that they carried out one of their most important scientific tasks. Using basic geological tools, Buzz Aldrin drove two core tubes into the lunar surface to collect the most famous rock samples taken in history. By the time they'd finished hammering and scooping up samples they had collected 22 kilogrammes (47 pounds) of lunar treasure. After using a pulley system to lift their scientifically priceless cargo on board, they closed the hatch and went to bed. As the two astronauts slept alongside the precious lunar rocks, the United States could justifiably claim to have won the greatest and arguably most glorious political victory in human history. For one rare moment, a political victory was also a triumph for all mankind.

However, it is not widely known that as the Apollo 11 lunar module rested on the Moon, a Soviet spacecraft was also in lunar orbit. The unmanned Luna 15 was the Soviets' third attempt to land on the Moon and collect lunar rock samples. Launched three days before Apollo 11, Luna 15 was a last-ditch attempt to win the scientific race to return rock samples from another world. Unfortunately, although Luna 15 successfully began its descent to the Moon's surface, it crashed into it shortly afterwards. Only Apollo 11 returned with moon rocks, which continue to be analysed to this day in the high-security labs of the lunar sample building in Houston, Texas.

Despite forty years of study, one thing has been clear pretty much from the start: these priceless examples of alien geology are remarkably similar to rocks found on Earth. In the main, they are composed of the common rock-forming elements oxygen, silicon, magnesium, iron, calcium and aluminium, but there is absolutely nothing on the Moon's surface that couldn't be found here on Earth.

Since Apollo 11's success, we have landed on Mars and Venus, parachuted into Jupiter's atmosphere, touched down on Saturn's moon Titan, and visited asteroids Eros and Itokawa and the comet Tempel 1. Each time the story is the same; the Solar System is made of the same stuff as we are. To date, eight landings on our nearest neighbour, Mars, have allowed us to explore the planet's geology in intimate detail. We now know Mars is rich in iron, which has oxidised to form its familiar rusty red colour, and that Martian soil is slightly alkaline and contains elements such as magnesium, sodium, potassium and also chloride. We also know that Venus' thick

TOP: On 21 July 1969, Neil Armstrong and Buzz Aldrin became the first humans to set foot on the Moon. This successful landing also opened up infinite possibilities for scientists to understand the formation of the lunar landscape. This photo shows Aldrin collecting some of the lunar rock samples that they took back to Earth for analysis.

ABOVE: The Apollo 11 lunar mission was launched from the Kennedy Space Center, Florida, on 16 July 1969 and safely returned to Earth on 24 July 1969, complete with its priceless cargo of samples from the Moon's surface. The first container was transferred to Ellington Air Force Base and was taken directly to the Lunar Receiving Laboratory at the Manned Spacecraft Center (MSC) in Houston, Texas.

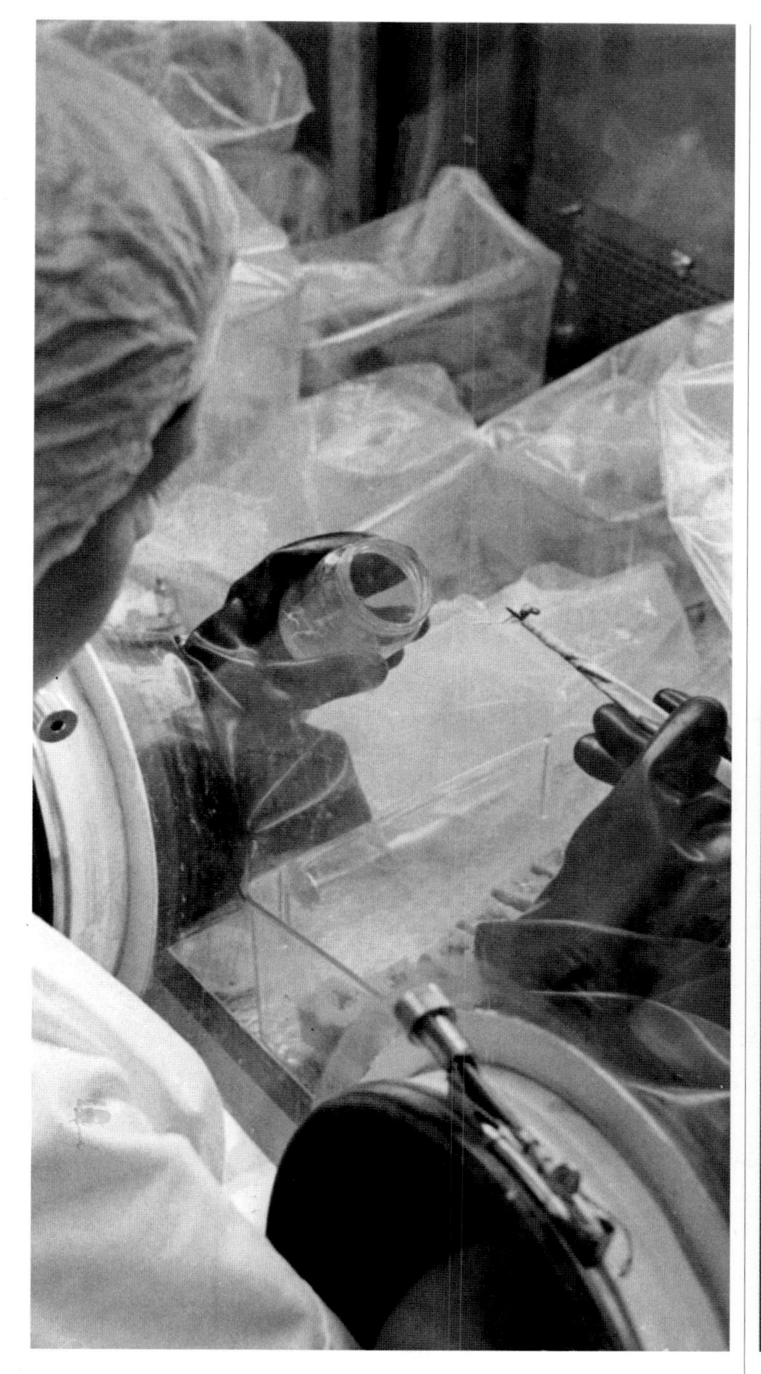

LEFT: Once safely returned to Earth, the treasures from the Moon, including rock samples, were painstakingly analysed at a high-security laboratory, and are still being used for analysis today.

BELOW: This false-colour photograph of Neptune was taken by Voyager 2. This image has enabled scientists to discover that the planet is rich in organic molecules such as methane.

Again and again we find there is much to discover in our solar system, but there are never new elements to unearth.

atmosphere is full of sulphur, and the planet Mercury is a large metal ball of iron with a thin crust comprised mostly of silicon. Even at the very edge of the Solar System, billions of miles away from Earth, we have discovered that Neptune is rich in organic molecules such as methane, a substance we find in abundance on our planet. Again and again we find there is much to discover in our solar system, but there are never new elements to unearth. From a scientific perspective this is unsurprising, because long ago Mendeleev's table revealed there isn't any room for other light elements in nature – we have discovered the full set. It would take a change in the laws of physics to discover something on the surface of another world that doesn't fit into Mendeleev's scheme, but from the explorer's perspective, seeing is believing!

So what about the rest of the Universe? How universal are these elements across the far reaches of the cosmos? Could it be that there are places in the distant Universe where the laws of physics are different? This is a legitimate question – we shouldn't simply assume that everything at the edge of the visible Universe, billions of light years away, operates exactly as it does here, no matter how persuasive the arguments from theoretical physics. Experiment and observation are the ultimate reality check. It may seem impossible to presume that we could ever answer this question directly and discover what the stars are made of, because they are so far away (they may indeed remain untouchable forever), but in fact we knew what the stars were made of long before we got our hands on that first piece of lunar rock ◉

WHAT ARE STARS MADE OF?

The Sun, the burning star at the heart of our solar system, is 150 million kilometres (93 million miles) away from Earth. Beyond that, the nearest known star, the red dwarf Proxima Centauri, requires a journey of over four light years or forty thousand billion kilometres (twenty-five thousand billion miles). We have learnt a lot about Proxima Centauri since it was discovered by Robert Innes at the Cape Observatory, in South Africa, in 1915. It is thought that Proxima Centauri is part of a triple star system with its neighbouring binary star system, Alpha Centauri A and B, and although it cannot be seen with the naked eye, we have been able to measure its mass and diameter and chart its brightness across the last 100 years. Despite the fact that our only contact with these neighbouring stars, and with any star other than our Sun, is the light that has crossed the Universe to reach us, we have been able to go much further than simply cataloguing their vital statistics. We can measure the precise constituents of any and every visible star in the sky, because encoded in the light that rains down on Earth is the key to understanding what they are made of. It is all made possible by a particularly beautiful property of the elements.

The tale of how we learnt to read the history of the stars in their light began with the work of Isaac Newton in 1670. In his 'Theory of Colour', Newton demonstrated that light is

ABOVE: Over a simple campfire I recreated the experiments of Gustav Kirchhoff and Robert Bunsen that made such a major impact in the development of quantum theory. Just as they discovered 150 years ago, when I threw the copper into the fire it burned with a spectacular blue flame.

RIGHT: In the early nineteenth century, German scientist Joseph von Fraunhofer documented the existence of 574 dark lines within the solar spectrum. This diagram is a visual representation of these Fraunhofer lines.

made up of a spectrum of colours, and that with nothing more complicated than a glass prism you can split the white light of the Sun into its colourful components. Almost 150 years later, the German scientist Joseph von Fraunhofer made a startling discovery about the solar spectrum whilst calibrating some of his state-of-the-art telescopic lenses and prisms. Lying within the solar spectrum, Fraunhofer documented the existence of 574 dark lines; there were literally hundreds of gaps – missing colours in the Sun's light. Unaware of the significance of this discovery at the time, Fraunhofer carefully mapped their positions in great detail. He went on to discover black lines in the light from the Moon and planets, and from other stars. These are now known as Fraunhofer lines.

Further work by two more of the great German scientists of the nineteenth century, Gustav Kirchhoff and Robert Bunsen (perhaps best known to schoolchildren everywhere as the inventor of the Bunsen burner), finally gave meaning to these lines. They surmised correctly that these black spectral lines were the fingerprints of the chemical elements in the atmosphere of the Sun itself. Across 150 million kilometres (93 million miles) of space, the light of our star had carried the signature of its constituents to us.

Kirchhoff and Bunsen's discovery was purely empirical – they had observed that when gases are heated on Earth they do not simply glow like a piece of hot metal, they give off light of very specific colours – and interestingly those colours depend only on the chemical composition of the gas and not on the temperature. In particular, each chemical element gives off its own unique set of colours. The element strontium, for example, burns with a beautiful red colour, sodium with a deep yellow, and copper is a haunting emerald green.

The two German scientists also noticed that the missing black lines in the solar spectrum corresponded exactly to the glowing colours of the elements. There are, for example, two black lines in the yellow part of the Sun's light that correspond exactly to the two distinct yellow emission lines of hot sodium vapour. You will be familiar with this mixture of two very slightly different yellows – it is the colour of sodium streetlights.

Interestingly, Kirchhoff and Bunsen had no idea why the elements behaved in this way, but this didn't matter if all you wanted to do was to match the signatures of elements observed on Earth with the signatures in the light from the Sun and stars. It wasn't until the turn of the twentieth century that an explanation for this strange behaviour of the elements was discovered. The answer lies in quantum mechanics, and the spectrographic work of physicists and chemists such as Kirchhoff and Bunsen was a major motivating factor in the development of the quantum theory. Elements emit and

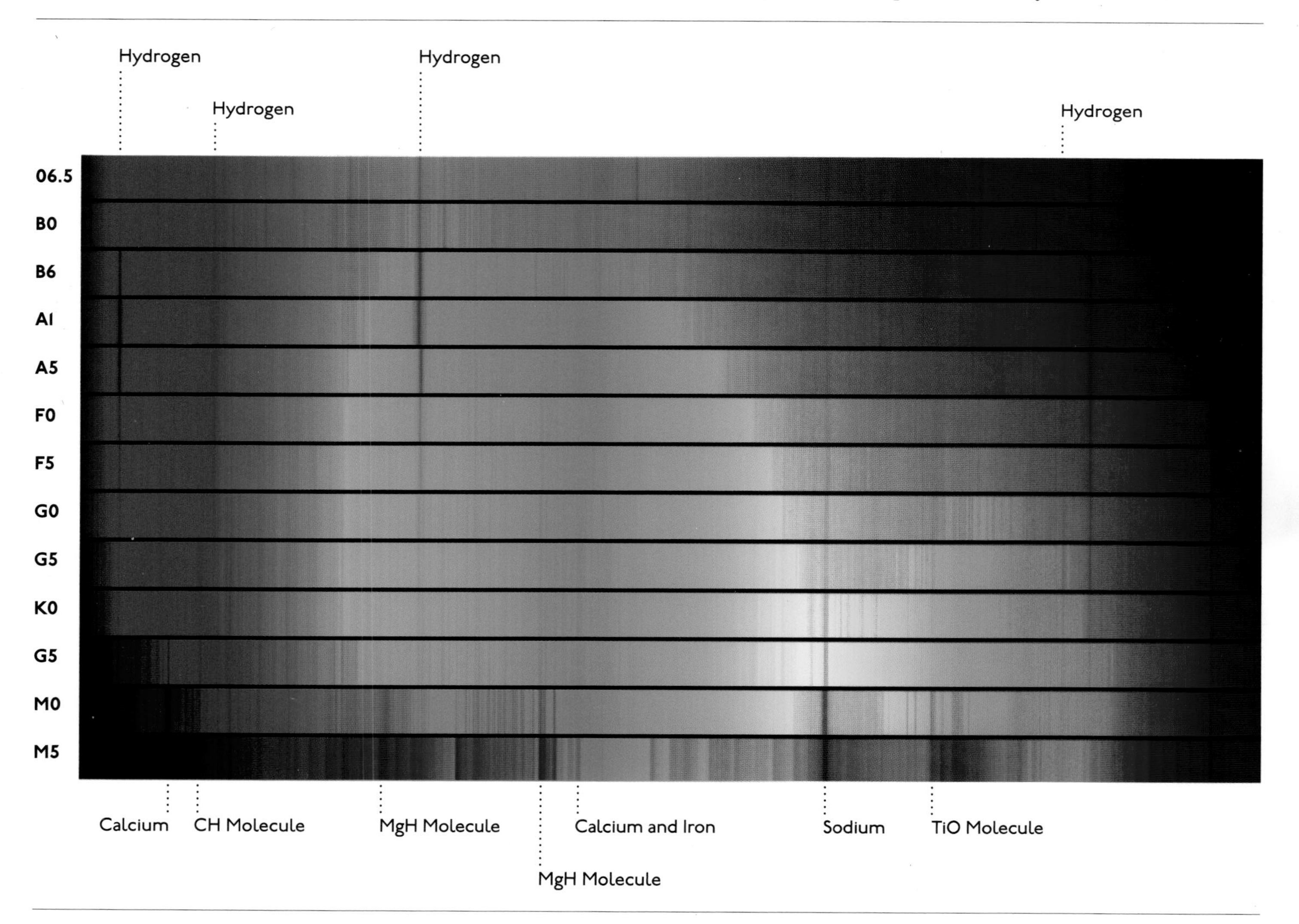

BELOW: Spectographic investigations have revealed that Sirius, the dog star, is metal-heavy, with an iron content three times that of the Sun.

RIGHT: Although Polaris, the pole star (top and middle), is 430 light years away, we know by looking that it has about the same heavy element abundance as our sun, but markedly less carbon and a lot more nitrogen. Vega (bottom), meanwhile, as the second-brightest star in the northern sky, consists of only about a third of the amount of metals as our sun.

Isn't it simply wonderful that just by looking at the light from those twinkling stars we can tell what those fiery worlds, so far away, are made of?

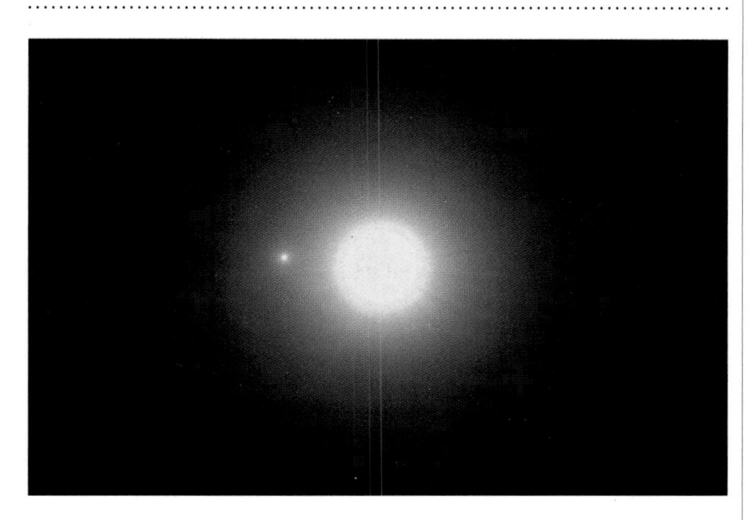

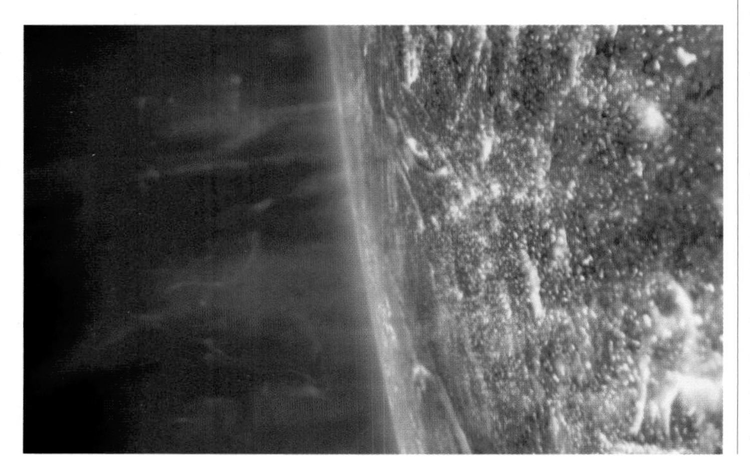

absorb light when the electrons surrounding their atomic nuclei jump around. The key insight that led to quantum theory was that electrons can't exist anywhere around a nucleus like planets around a star, but they are instead placed in specific, very restrictive 'orbits'. The deep reason for this is that electrons do not always behave as point-like particles of matter. They also exhibit wave-like properties, and this severely restricts the ways in which they can be confined around the atomic nucleus. What happens at a microscopic level when an atom absorbs some light is that an electron jumps to a different, more energetic, orbit and it emits light when the electron falls back from a higher to a lower energy orbit. The difference in energy between the lower orbit and the higher orbit must correspond exactly to the energy of the light absorbed or emitted.

However, quantum theory also stipulates that light should not always be thought of as a wave. Just like electrons, light can behave as both a wave and a stream of particles. These particles are called photons. Now, here is the key point: photons of a particular energy correspond to a particular colour of light, so red photons have a lower energy than yellow photons, which have a lower energy than blue photons. Since each element has electrons in unique orbits around the nucleus, this means that each element will only be able to absorb particular photons in order to move its electrons around into higher energy orbits. Conversely, when the electrons drop from higher to lower energy orbits, they will only emit photons of a particular energy and therefore a very particular colour. This is what we see when we observe the elements emitting or absorbing particular colours of light. We are in a very real sense seeing the structure of the atoms themselves.

When looking at a spectrum of light from our sun you can see hundreds of Fraunhofer lines, and each and every one of those corresponds to a different element in the solar atmosphere which absorbs light as it passes through. From sodium in the yellow, through iron, magnesium, and all the way across to the so-called hydrogen alpha line in the red, the signatures of each of the elements are encrypted in the solar code.

So by looking at these lines in precise detail you can work out exactly which elements are present in the Sun. This turns out to be roughly 70 per cent hydrogen, 28 per cent helium, and the remaining 2 per cent is made up of the other elements.

It is worth repeating here that you can apply this theory not only to the Sun, but for any of the stars you can see in the sky – which allows us to measure the constituents of their atmospheres with extraordinary accuracy. Isn't it simply wonderful that just by looking at the light from those twinkling stars we can tell what those fiery worlds, so far away, are made of?

These spectrographic investigations of the light from the cosmos have confirmed what our scientific intuition suggested to us: wherever we look, we only ever see the signatures of the set of ninety-four naturally occurring elements that we have collected and identified here on Earth.

So it is clear that we are connected in a very real sense to the whole of the Universe – with its hundreds of billions of stars across billions of galaxies – because we are all intrinsically made of the same stuff. And, as we will explain, there is one very simple reason for that: everything in the Universe shares the same origin ◉

THE EARLY UNIVERSE

In order to understand where we come from we have to understand events that happened in the first few seconds of the life of the Universe. When the Universe began it was unimaginably hot and dense – we literally don't have the scientific language to describe it. It was beautiful in a very real sense. There was no structure, there was certainly no matter, and it was exactly the same whichever way you looked at it. It's a difficult concept to grasp, but we can get some idea of what happened to the early Universe by looking at the behaviour of one of the most common substances on Earth: water.

RIGHT: Water is one of the most common substances on Earth, but it can produce some of the most spectacular geological wonders on our planet. The El Tatio Geysers in Chile are just one example of water's awesome activity.

BELOW: One of Earth's most incredible natural wonders, the El Tatio Geysers make up the largest geyser field in the world. As they are located at a height of 4,200 metres (13,800 feet) in the Chilean Andes, they are also the highest.

EL TATIO GEYSERS, CHILE

High in the Andes Mountains, in the far north of Chile, you will find the spectacular El Tatio Geysers. Erupting at 4,200 metres (13,800 feet) above sea level, this is one of the geological wonders of Earth's Southern Hemisphere. Not only is it one of the largest geyser fields in the world, it is also one of the highest. For those who journey here to witness the eruption of the jets of water skywards there is only one time to visit – sunrise.

In the early morning, as the Sun begins to peer over the horizon, the combination of super-heated water and freezing cold air produces a rare phenomenon. Like all geysers, the boiling water delivered to the surface by the geological plumbing bursts out and flashes into steam, forming the majestic columns. But here, because of the high altitude and bitter temperatures, the steam rapidly condenses and returns to its frozen state, covering the ground with sheets of ice. It is surely one of the most spectacular naturally occurring locations on the planet in which you can see water in all three of its phases: liquid, vapour and solid ice. It is this rapid transformation of water through its three familiar phases that provides us with a nice analogy to discuss events that happened in the very early life of the Universe.

A water molecule is made up of two chemical elements: oxygen and hydrogen. Oxygen and hydrogen atoms are symmetric when they are alone and uncombined. This particular use of the word symmetric is perhaps unfamiliar; what is meant in this context is that the atoms themselves would look the same no matter what angle you viewed them from. In the language of physics, this is called rotational symmetry. A perfect sphere has perfect rotational symmetry, because whichever way you look at it or spin it around it looks exactly the same. When an oxygen atom combines with two hydrogen atoms to form a water molecule – H_2O – this rotational symmetry disappears because the water molecules have a particular shape – there is an angle of 105 degrees

BELOW: Approximately 70 per cent of Earth's surface is covered by water. At the El Tatio Geysers you can see water in all its three forms.

Walking through pools of water on the ground, I held a sheet of glass in the geysers' steam and watched ice crystals form on it.

Exactly like the journey of steam to ice, of chaos to order, this was the Universe in transition. A transition where the structure and substance of all the particles of matter emerged for the first time.

between the hydrogen and oxygen atoms. A physicist would say that the symmetry is now broken, because the water molecule has a distinct orientation. We can break the symmetry of water still further by cooling down all the molecules until they stick together and solidify into ice. Now the crystals of ice are beautiful and almost impossibly intricate; full of structure and a complexity that completely hides the perfect symmetry of the original atoms, and also the simple but different symmetry of the water molecules themselves.

The important point here is that all this complexity emerged when the symmetry was broken, but we did nothing to the water itself to break its symmetry other than cool it down. So although it looks for all the world as if a master sculptor sat down and chiselled out beautiful patterns in the ice, this intricacy and beauty emerged completely spontaneously out of building blocks that are themselves utterly symmetric.

Physicists call this process spontaneous symmetry breaking, and it is this idea that lies at the heart of our understanding of the early Universe ◉

THE BIG BANG

Thirteen billion years ago the Universe began in the event called the Big Bang. We don't know why. We also don't know why it took the initial form that it did. This is one of the unsolved mysteries that makes fundamental physics so exciting. The first milestone we can speak of in anything resembling scientific language is known as the Planck Era, a period that occurred a mind-blowing 10^{-43} seconds after the Big Bang. When written in full, that number has 42 decimal places: 0.0000000000000000000000000000000000000000001 seconds. That's not very long at all. This number can be arrived at very simply because it is related to the strength of the gravitational force. It is so incredibly tiny ultimately because gravity is so weak – and we don't know the reason for that, either! At that time the four fundamental forces of nature that we know of today – gravity, the strong and weak nuclear forces, and electromagnetism – were one and the same force, a single 'superforce'. There was no matter at this stage, only energy and the superforce. This is what a physicist would call a very symmetric situation.

As the Universe rapidly expanded and cooled it underwent a series of symmetry-breaking events. The first, at the end of the Planck Era, saw gravity separate from the other forces of nature, and so the perfect symmetry was broken. Around 10^{-36} seconds after the Big Bang, another symmetry-breaking event occurred which marked the end of the Grand Unification Era. This saw the strong nuclear force (the force that sticks the quarks together inside protons and neutrons) split from the other forces. At this point the Universe underwent an astonishingly violent expansion known as inflation, in which the Universe expanded in size by a factor of 10^{26} (that's 100 million million million million times) in an unimaginably small space of time – it was all over in 10^{-32} seconds. This was when sub-atomic particles entered the Universe for the first time, but they weren't quite what we see today because none of them had any mass at all.

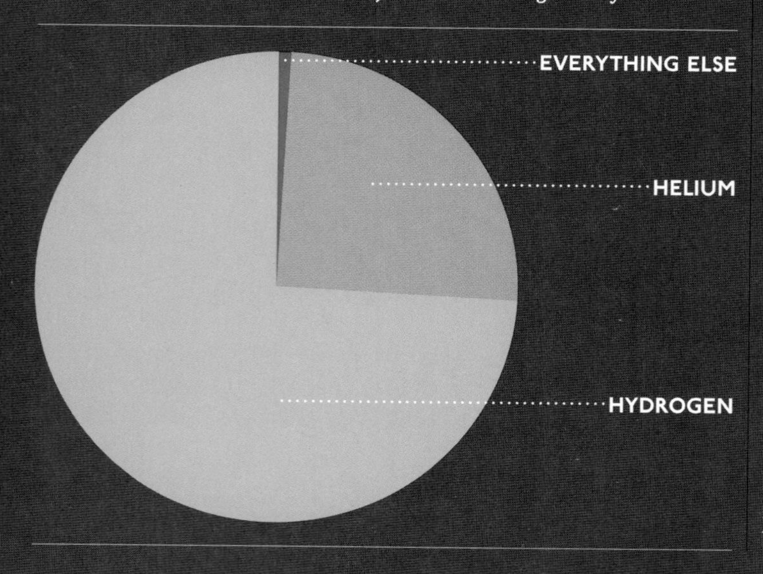

Up until this point this story is theoretically well-motivated but experimentally relatively untested. The next great symmetry-breaking event, however, which occurred 10^{-11} seconds after the Big Bang, is absolutely within our reach because this is the era we are recreating and observing at CERN's Large Hadron Collider. It is called electroweak symmetry breaking; at this point the final two forces of nature – electromagnetism and the weak nuclear force – are separated. During this process the sub-atomic building blocks of everything we see today (the quarks and electrons) acquired mass. The most popular theory for this process is known as the Higgs mechanism, and the search for the associated Higgs Particle is one of the key goals of the Large Hadron Collider project.

We are now on very firm experimental and theoretical ground. From this point on we know pretty much exactly what happened in the Universe because we can do experiments at particle accelerators to check that we understand the physics. The emergence of the familiar particles and forces we see in the Universe today happened, we believe, as a result of a series of symmetry-breaking events which began way back at the end of the Planck Era. The concept of spontaneous symmetry breaking in the early Universe is exactly the same as for the transitions from water vapour to liquid water to ice. Complex patterns emerge without prompting – just as a result of falling temperature – and these patterns obscure the underlying symmetry of the initial state. So just as the seemingly infinite complexity of snowflakes masks the simple symmetry of oxygen and hydrogen atoms, so the array of forces of nature and sub-atomic particles we see as the building blocks of the Universe today obscures the symmetry of the early Universe.

There is now one final step needed to arrive at the protons and neutrons – the building blocks of the elements – and the first elements themselves. This began around a millionth of a second after the Big Bang, when the quarks had cooled enough to become glued together by the strong nuclear force to form protons and neutrons. The simplest element, hydrogen, consists of a single proton. So after only a millionth of a second in the life of the Universe, the first chemical element had made an appearance. After three minutes, the Universe was cold enough for the protons and neutrons themselves to stick together to form helium. With two protons and one or two neutrons in its nucleus, helium is the second-simplest chemical element. There were also very, very small amounts of lithium, with three protons, and beryllium, with four protons – the third- and fourth-simplest elements. And this is pretty much where the process stopped. After three minutes the Universe had the four distinct forces we know of today – gravity, the strong and weak nuclear forces, and electromagnetism, and was composed of roughly 75 per cent hydrogen (by mass) and 25 per cent helium. This is the story of the creation of the simplest chemical elements and of successive symmetry-breaking events in the early Universe ◉

LEFT: Careful scientific study leads us to conclude that the building blocks of our Universe are fundamentally hydrogen and helium.

BELOW: A computer simulation of an event showing the decay of Higgs Bosen producing four muons (white tracks). This image shows how the Higgs Bosen might be seen in the CMS detector from the Large Hadron Collider at CERN.

SUB-ATOMIC PARTICLES

Our understanding of the structure of matter has increased in the last century. Originally, atoms were thought to be the basic building blocks of life, but Rutherford's famous diffraction experiment proved that matter consisted mainly of space, with each atom containing a very small dense nucleus surrounded by a cloud of electrons. Further investigation showed that each nucleus was composed of protons and neutrons and that each proton was composed of up and down quarks. We have now reached what is believed to be the smallest particles possible – scientists have now discovered that all matter is composed of 9 particles and 4 forces, plus the hypothetical Higgs Boson. The search for the basic building blocks of matter has used matter colliders, which can produce the very high energies that are required to recreate the temperatures in the early Universe, when these sub-atomic particles originally existed.

e⁻

ELECTRON

PHOTON

NUCLEUS

e⁻

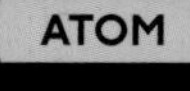

Electrons (e⁻) are bound to nucleus by electromagnetic force, which is mediated by photons

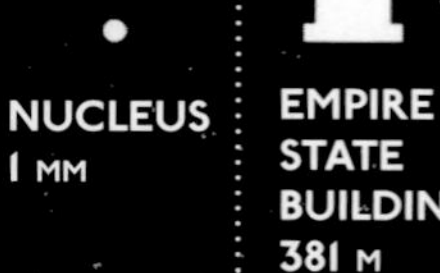

ATOM

Protons and neutrons are bound together in the nucleus by the strong force which is mediated by gluons

NEUTRON

PROTON

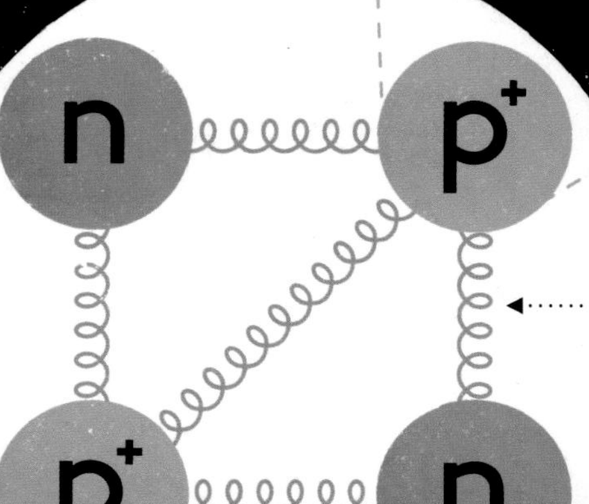

GLUON

UP QUARK

DOWN QUARK

d

u

u

GLUON

PROTON

The proton is made up of two 'up' quarks and one ''down' quark. They are bound together inside the proton by the strong force (mediated by gluons)

PARTICLE CONTENT OF THE STANDARD MODEL

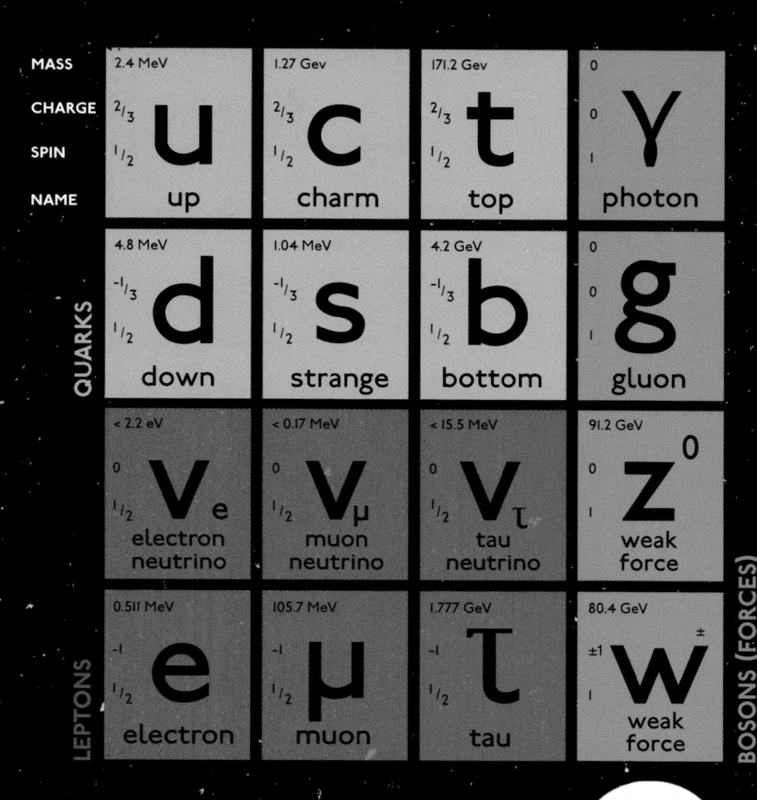

H
Higgs
Boson

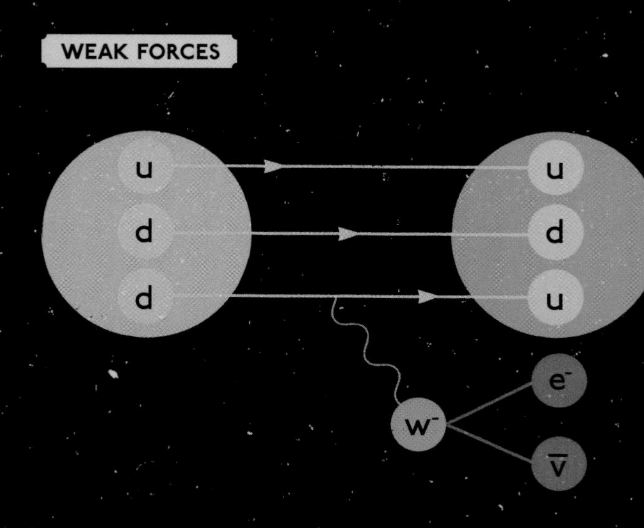

The weak force converts protons into neutrons, the first step in the nuclear fusion process in the Sun.

TIMELINE OF THE UNIVERSE: THE BIG BANG TO THE PRESENT

The history of the Universe can be split into several phases, according to the physical conditions that existed at the time. Things happened quickly in the first fractions of a second, when the Universe was filled with an intensely hot soup of energy and exotic particles. From this emerged the first protons and neutrons which were later to form the nuclei of the first atoms – mostly hydrogen and helium. After the emission of the cosmic microwave background, around 400,000 years after the Big Bang, the pace of events became more sedate. According to current understanding, the Universe will continue to expand forever, eventually fading into darkness in the unimaginably distant future.

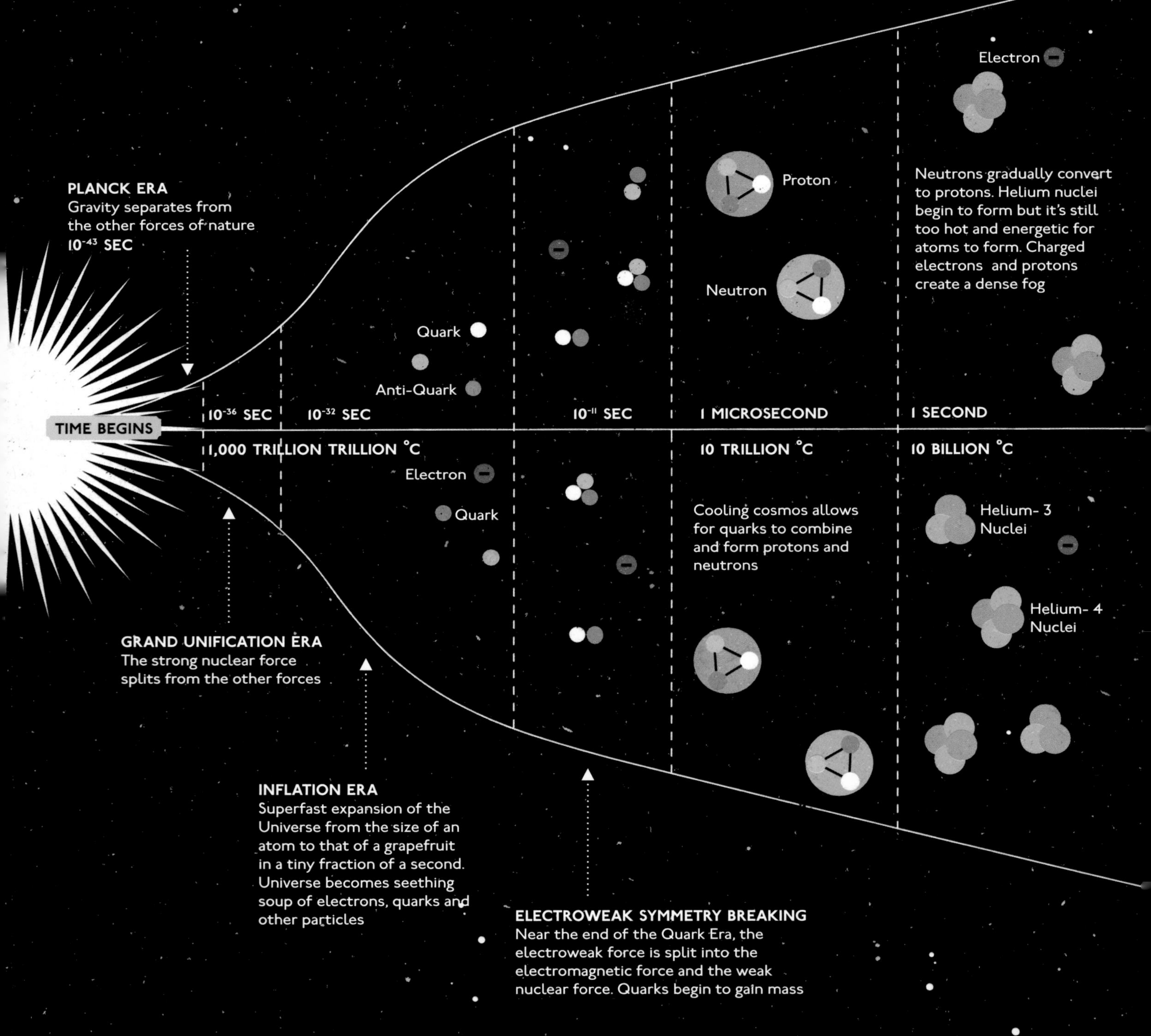

TEMPERATURE COOLING

Gravity makes hydrogen and helium gas coalesce to form giant clouds that will later become galaxies. Smaller clumps of gas collapse to form the first stars

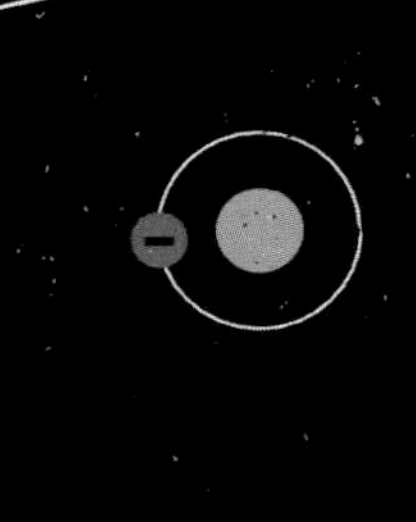

Hydrogen Atom

Helium Atom

400,000 YEARS | 1 BILLION YEARS | 13.7 BILLION YEARS | PRESENT DAY

2,700 °C | -200 °C | -270 °C

Electrons combine with protons and neutrons to form atoms, mostly Hydrogen and Helium, Light can begin to shine

Galaxies cluster together under gravity. The first stars die and spew heavy elements into space. These will eventually form new stars and planets

SPACE EXPANDING

There's a mystery at the heart of science for which, as yet, we have no explanation, and that is that this universe is simple. Underlying all of the astonishing complexity appears to be a magnificent simplicity, and nowhere is that simplicity more obvious than in the construction of the elements.

LEFT: The construction of all the chemical elements in the Universe can be illustrated with the most basic demonstration – so simple, it's child's play. To understand how the structure has emerged, all you need is a pot of bubble mixture. Blow one bubble and you have returned to the beginning of time, when all that existed in the Universe was the proton.

MATTER BY NUMBERS

Throughout human history the discovery and use of specific chemical elements has been intricately linked with the rise of civilisation. It is believed that copper was first mined and crafted by humans 11,000 years ago, and the specific characteristics of this metal ushered in a new age of technology and the transition from stone tools and weapons to metal ones. Four thousand years later it happened again but with iron which, even today, when mixed with carbon to form the alloy steel is the exoskeleton of industrial civilisation.

These two elements played a role in our history because of their particular physical characteristics. Copper was almost certainly the first metal to be used by humans; as it is such an unreactive chemical that it is one of the few metals that occurs naturally in its pure state. It is also very soft and malleable and so relatively easy to work into tools and weapons. When combined with another metallic element – tin – copper forms the alloy bronze; when combined with zinc it forms brass. Iron is, perhaps surprisingly, the most abundant element on Earth, and the fourth-most abundant element in the rocks of Earth's crust. Although more difficult to extract and work with than bronze, iron is an excellent material for weapons manufacture as it is harder and lasts longer than bronze.

These two metals have had a profound influence on human history and sit just a couple of spaces apart in the periodic table. Iron (Fe) is element number 26 and copper (Cu) is at 29. The first humans to use these metals would, of course, have had no idea of the reason for the physical similarities and differences between the two elements. So what is the fundamental difference between them? The answer is remarkably simple. As described earlier, the atoms of each element are composed of three building blocks: protons, neutrons and electrons. We do not need to consider the quarks inside the protons and neutrons, because at the temperatures we encounter on Earth they stay locked away. So when discussing Earthly chemistry, we can ignore them.

We have already encountered the first four elements; one of these, hydrogen, has an atomic nucleus consisting of a single proton. The proton has a positive electric charge, which allows it to trap an electron in orbit around it to form a hydrogen atom. The electron carries a negative electric charge, equal and opposite to that of the proton. This means that hydrogen atoms are electrically neutral. The reason why the electron has exactly the equal and opposite charge of the proton is not known. This is even more surprising when you look at the quarks that build up the proton. The proton is made up of three quarks – two up quarks and one down quark. The up quark has an electric charge of +2/3, and the down quark has a charge of -1/3. The electron has a charge of -1. So it is only when they are combined to form a proton that everything balances out properly. The neutron consists of two down quarks and an up, which means that it has no electric charge at all. This cannot be a coincidence, and it is one of the great challenges for twenty-first-century physics to explain it.

Chemical elements differ because of varying numbers of protons in their atomic nucleus, but the number of neutrons makes no difference to their chemical properties. Chemistry is down to the way the electrons behave that orbit around the nucleus, and the number of electrons is equal to the number of protons. As we know, the hydrogen atom consists of one proton and one electron, but there is another form of hydrogen called deuterium. Deuterium has a neutron attached to the proton inside its nucleus, but this doesn't change its chemical properties as there is still only one electron. Technically speaking, deuterium and hydrogen are two different isotopes of the same element. Helium atoms always have two protons and two electrons; it also has forms with one and two neutrons, known as helium-3 and helium-4 respectively. Next comes lithium, with three protons, three electrons and either three or four neutrons, sometimes more. Carbon has six protons and varying numbers of neutrons, and so on. The rule is that each successive element has one more proton in its nucleus, and at least one more neutron, although the number of neutrons varies. The neutrons help the nucleus to stick together; which is bound tightly by the strong nuclear force, and neutrons add to this, even though they have no electric charge. Electric charge is a bad thing for the nucleus; because the protons are positively charged, they repel each other and try to blow the nucleus apart. The neutrons don't suffer from this problem, which is one of the reasons why heavier nuclei tend to have more neutrons than protons.

So the construction of chemical elements is simple. If you want to turn iron into copper, add three protons and a handful of neutrons to its nucleus. That's all there is to it. This is easier said than done, of course, yet nature can do it because when the Universe was only a few minutes old the first four chemical elements existed. The building blocks were present, but the heavier elements were assembled later ◉

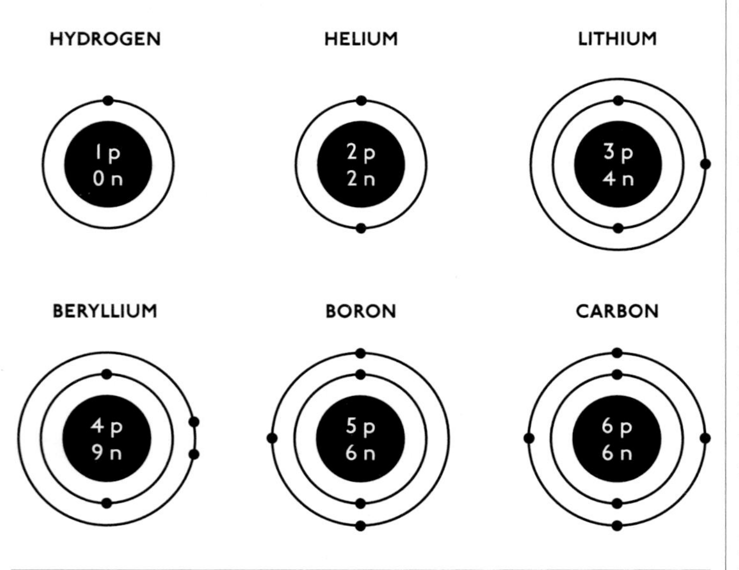

THE SIX LIGHTEST ELEMENTS IN NATURE: HYDROGEN TO CARBON
In each element the number of protons (p) in its nucleus is the same as the number of orbiting electrons, but the number of neutrons (n), which have no electric charge, can vary.

THE MOST POWERFUL EXPLOSION ON EARTH

BELOW: The now iconic image of a hydrogen bomb explosion. This mushroom cloud was produced by the detonation of XX-33 Romeo on 26 March 1954; it was the third-largest test ever detonated by the USA.

Years before the Manhattan project designed and delivered the most destructive weapon used in anger in the history of warfare, two of the greatest physicists of the age had already lost interest in the idea. Edward Teller and Enrico Fermi were friends and colleagues who would both go on to be members of the Manhattan team, but in 1941, before any type of nuclear bomb had been assembled, their minds were already wandering beyond the bomb that would later be dropped on Hiroshima and Nagasaki with devastating effect.

The Hiroshima and Nagasaki bombs were fission bombs, which work by splitting the nuclei of very heavy elements (uranium in the case of the Hiroshima bomb and plutonium for the Nagasaki bomb), into lighter elements such as strontium and caesium. This is the assembly of the elements in reverse. Each time a nucleus of uranium or plutonium splits, neutrons are released which trigger the splitting of other nuclei. In this way a nuclear chain reaction ensues. Each time a heavy nucleus splits, a large amount of energy is liberated – this 'nuclear binding energy' is stored in the strong nuclear force field that sticks the protons and neutrons together inside the nucleus.

However, even in the very early stages of the Manhattan project, years before the idea of a fission bomb was a physical reality, Enrico Fermi postulated that there was the very real possibility of creating a far more powerful type of bomb. Edward Teller became obsessed with his friend's idea and spent the next decade designing and building a device that would create the most powerful explosions ever made on Earth. It earned Teller the title 'father of the hydrogen bomb'.

On 1 November 1952, the fruits of Fermi's conversation with Teller were realised. Ivy Mike was the codename given to the first successful testing of a hydrogen bomb on Enewetak, an atoll in the Pacific Ocean. The explosion was estimated to be 450 times more powerful than the bomb dropped on Nagasaki, producing a fireball over five kilometres (three miles) wide, a crater two kilometres (one mile) wide and wiping the tiny atoll off the map. Teller had collaborated with another Manhattan scientist, Stanislaw Ulam, to design the bomb, but he wasn't present for the explosion. Instead he sat watching a seismometer thousands of miles away in his office in Berkeley, California. The explosion was so powerful that he was able to clearly see the shockwave from the comfort of his office. 'It's a boy!', he cryptically told his colleagues to inform them of the success.

The Ivy Mike test was the first man-made nuclear fusion reaction. Nuclear fusion is the direct opposite of fission; it is the process by which two atomic nuclei are fused to form a single heavier element. The hydrogen bomb reproduces the process that occurred in the first seconds of the evolution of the Universe – the assembly of hydrogen into helium.

The Teller–Ulam design for the hydrogen bomb that exploded on Enewetak is the basic design employed by all five of the major nuclear weapon states today. Although the fusion element of the design is only part of its explosive power, combined with the other stages contained within the bomb it creates destruction on an unparalleled scale.

Here are two completely different ways of creating new elements and releasing vast amounts of energy. The first, fission, involves taking a heavy element and splitting it. The second, fusion, involves taking lighter elements and sticking

Look up into a clear blue sky and you are bathing in the energy of nuclear explosions on an unimaginable scale.

them together. But how can both these processes result in energy being released? Isn't there a contradiction here? There isn't, of course, because this is how nature works. It's all down to the delicate balance between the electric repulsion of the protons in the nucleus and the power of the strong nuclear force to stick the protons and neutrons together. Since there are two competing forces, one trying to blow the nucleus apart and one trying to glue it together, you might think there must be some kind of balancing point – an ideal mixture of protons and neutrons that is perfectly poised between attraction and repulsion. There are in fact two elements that are very close to the mixture of optimal stability, and these are iron and nickel. Elements lighter than these can be made more stable, releasing energy in the process, by fusing them together. Elements heavier than these can be made more stable, releasing energy in the process, by breaking them apart.

To be completely accurate, we should mention that there are other factors than just the balance between the electromagnetic and nuclear forces that feed into the stability of the elements. These are to do with the shape of the nucleus itself and that the balance between protons and neutrons is favoured for quantum mechanical reasons. (If you are interested, google 'Semi-empirical mass formula' and enjoy!)

Here on Earth, fusion may seem the ultimate human technological achievement but actually it's the most natural thing in the world. It didn't only happen at the Big Bang; it's a process that can be found occurring across the Universe as we speak. In fact, it illuminates the whole Universe and happens all the time millions of miles above our heads.

Fusion is the process that powers every star in the heavens, including our sun. Look up into a clear blue sky and you are bathing in the energy of nuclear explosions on an unimaginable scale. Deep in the Sun's core, 800,000 kilometres (500,000 miles) below the surface (where temperatures reach

BELOW: The shining Sun is one of the most natural demonstrations of the effect of fusion. It, and all the other stars in the heavens, are powered by the fusing of hydrogen and helium.

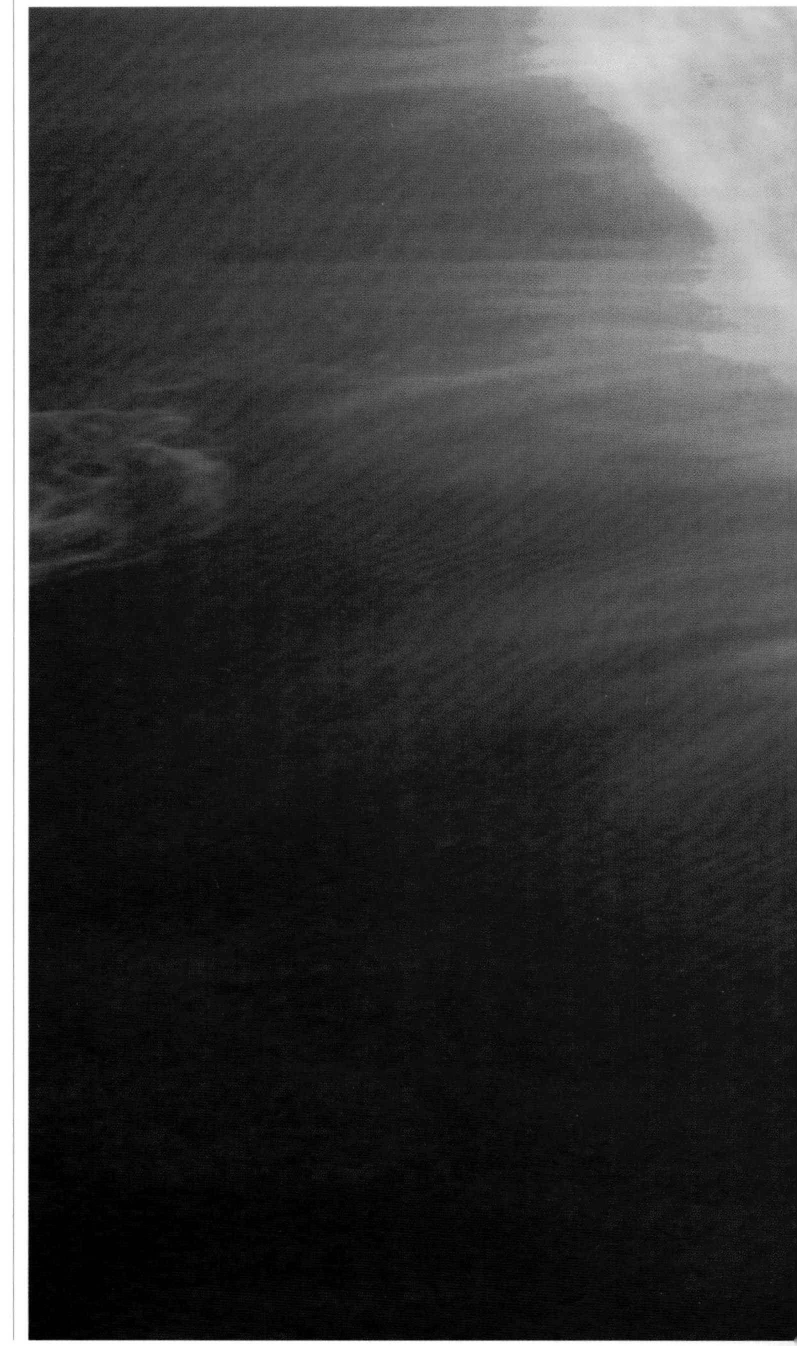

fifteen million degrees Celsius), the Sun is busy fusing hydrogen into helium at a furious rate. In just one second the Sun converts 600 million tonnes of hydrogen into helium, releasing as much energy as the human race will use in the next million years. This is the energy that makes the stars shine and fills the Solar System with heat and light.

It is the process of turning hydrogen into helium that creates the energy that allows all life on Earth to exist, but for all its power the Sun only converts hydrogen, the simplest element, into helium, the next simplest. This process is repeated across the night sky; every star in the Universe began its life fuelled by hydrogen and powered by this reaction.

So the assembly of the second-simplest element, helium, is well understood. We know the stars can do it, we know it happened in the very early Universe, and we can even do it ourselves on Earth. But this doesn't help to explain the origin of the other ninety-two naturally occurring elements. Clearly, somewhere in the Universe there must be a plentiful source of the other elements because they are everywhere, our whole planet is made from them. We are made of billions and billions of atoms; from magnesium, to zinc, to iron and, of course, the one atom that life is more dependent on than any other – carbon. Every human being on the planet is made from about a billion billion billion carbon atoms. That's an unimaginable number of carbon atoms that simply didn't exist in the early moments of the Universe. Where did they come from? The answer must be nuclear fusion, and the natural place to look is within the stars themselves ◉

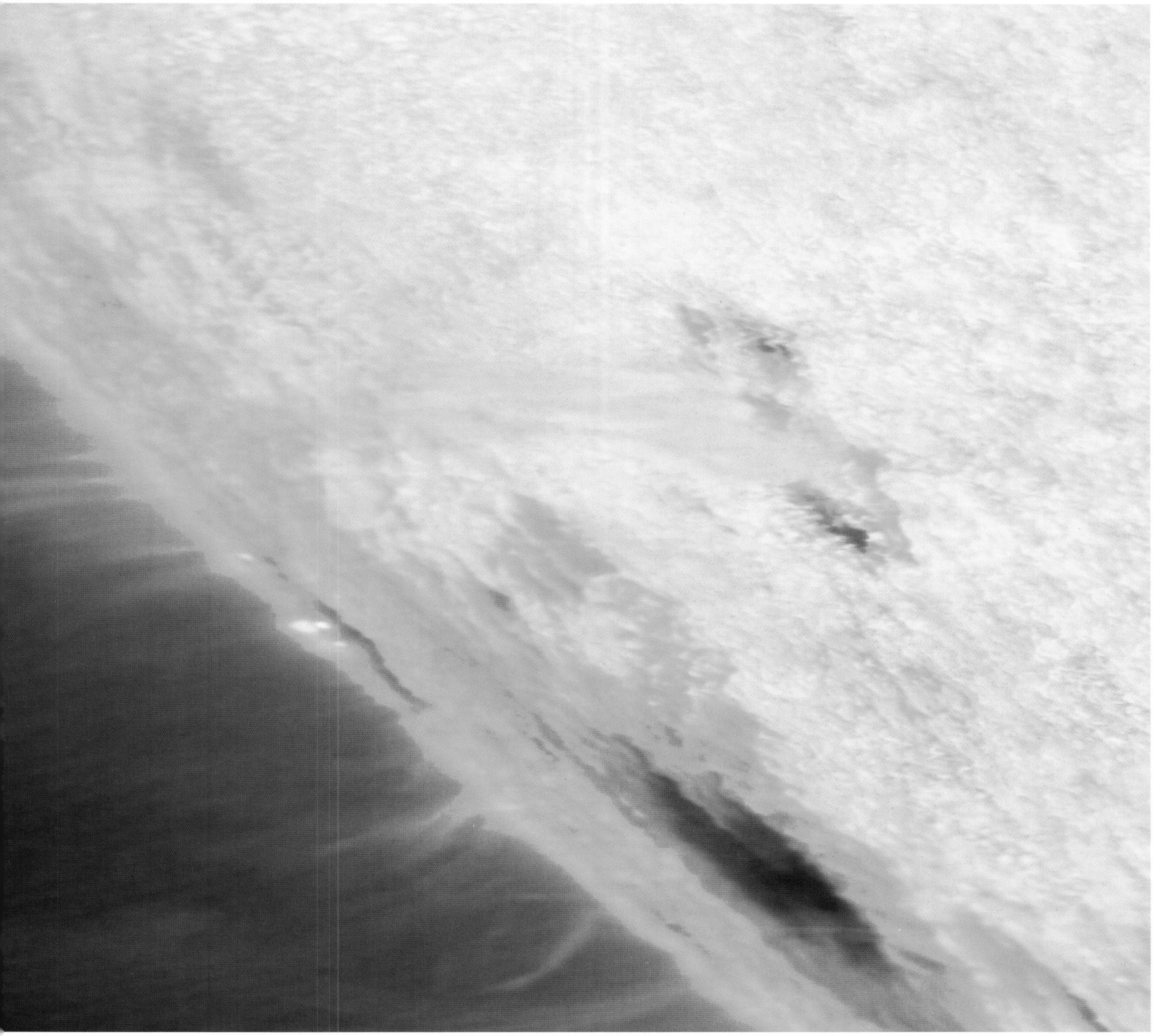

FROM BIG BANG TO SUNSHINE: THE FIRST STARS

The first stars formed around 100 million years after the Big Bang. The rate at which they burned their hydrogen fuel essentially depends on their mass. The more massive the star, the brighter it shines and the shorter its lifetime. The key to understanding how the heavier elements came into being lies in what happens to stars when they have exhausted their hydrogen fuel. For the most massive known stars, this may take only a few million years. For stars like our sun, it may take ten billion years – but the Universe has been around for plenty of time to allow generations of stars to live and die.

RIGHT: The brightly shining constellation of Orion is clearly visible as it sets in the night sky.

RED GIANT

As a star exhausts its hydrogen stores you might expect it to slowly flicker away, but for stars like our sun, the opposite happens. Having spent millions or billions of years with the core as its beating heart, a star that is running out of hydrogen in fact swells up to potentially hundreds of times its original size. Such stars are known as red giants.

One of the closest red giants to Earth is the star Alpha Orionis, better known as Betelgeuse, the ninth-brightest star in our night sky and one of our nearest neighbours in cosmic terms, a mere 500 light years away. Betelgeuse has long been familiar to stargazers, notable for its brightness and reddish tinge that is clearly visible to the naked eye. Sir John Herschel studied the star intensely in the nineteenth century, recording the dramatic variations in its brightness. However, it was only when three astronomers from the Mount Wilson Observatory in California tried to measure its diameter that we realised this was no ordinary star. Albert Michelson, Francis Pease

Betelgeuse is a vast wonder that would fill our solar system with a single wispy star.

and John Anderson used a specially designed telescope to measure the scale of this red star using a technique known as interferometry. By measuring the angular diameter (the apparent size of an object from our position on Earth), they came up with a number that, although it's been refined since, revealed something profound: Betelgeuse is a true giant in every sense. This star is about twenty times the mass of our sun but its size is rather more impressive. If you put Betelgeuse at the centre of our solar system it would dwarf our sun. In fact, Betelgeuse would extend past the Earth's orbit, encompassing everything out to Jupiter. Current estimates suggest it is around 800 million kilometres (500 million miles) in diameter; a vast, ethereal wonder that would fill our solar system with a single wispy star.

Due to its immense size and relative proximity, we can study Betelgeuse in incredible detail. In 1996, the Hubble Space Telescope took a picture of Betelgeuse that was the first direct image of another star to reveal its disc and surface features. We've even imaged sunspots on its surface and been able to study its atmosphere in ever-increasing detail. However, it's not the surface of the red giant that holds the clue to where the heavy elements are made; to understand that, we need to journey deep into its dying heart ◉

BELOW: These images of Betelgeuse are based on pictures taken by the Very Large Telescope at the European Southern Observatory, in Chile, and show gas plumes bursting from the star's surface into space.

RIGHT: Betelgeuse is the ninth-brightest star in our galaxy and one of our nearest neighbours. It can be seen from Earth with the naked eye – easily identifiable in the night sky for its brightness and reddish tinge.

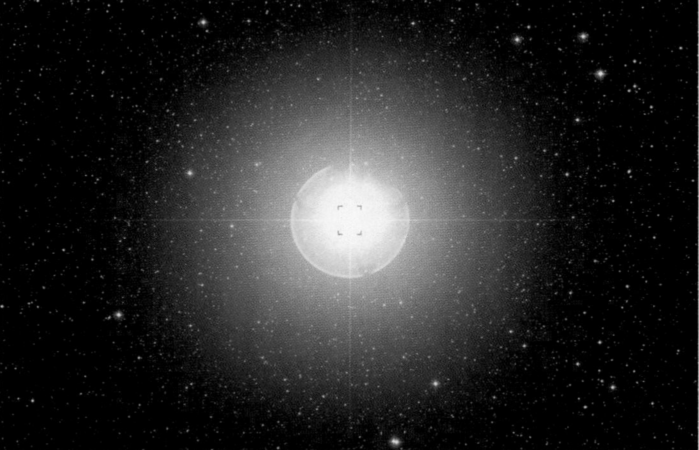

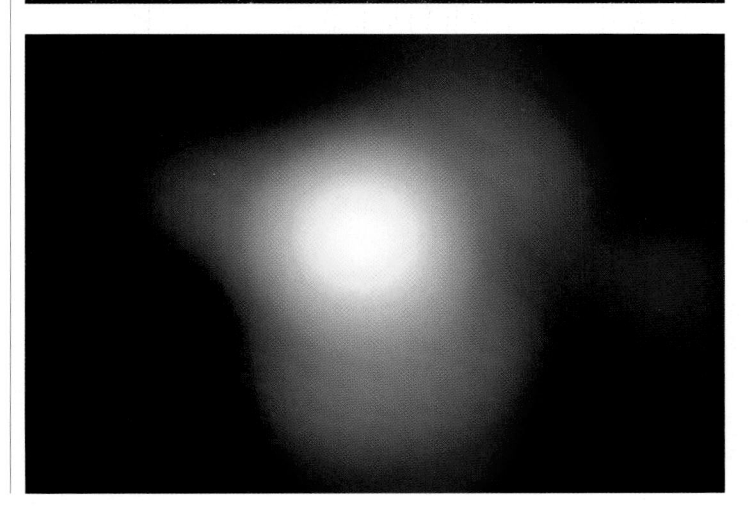

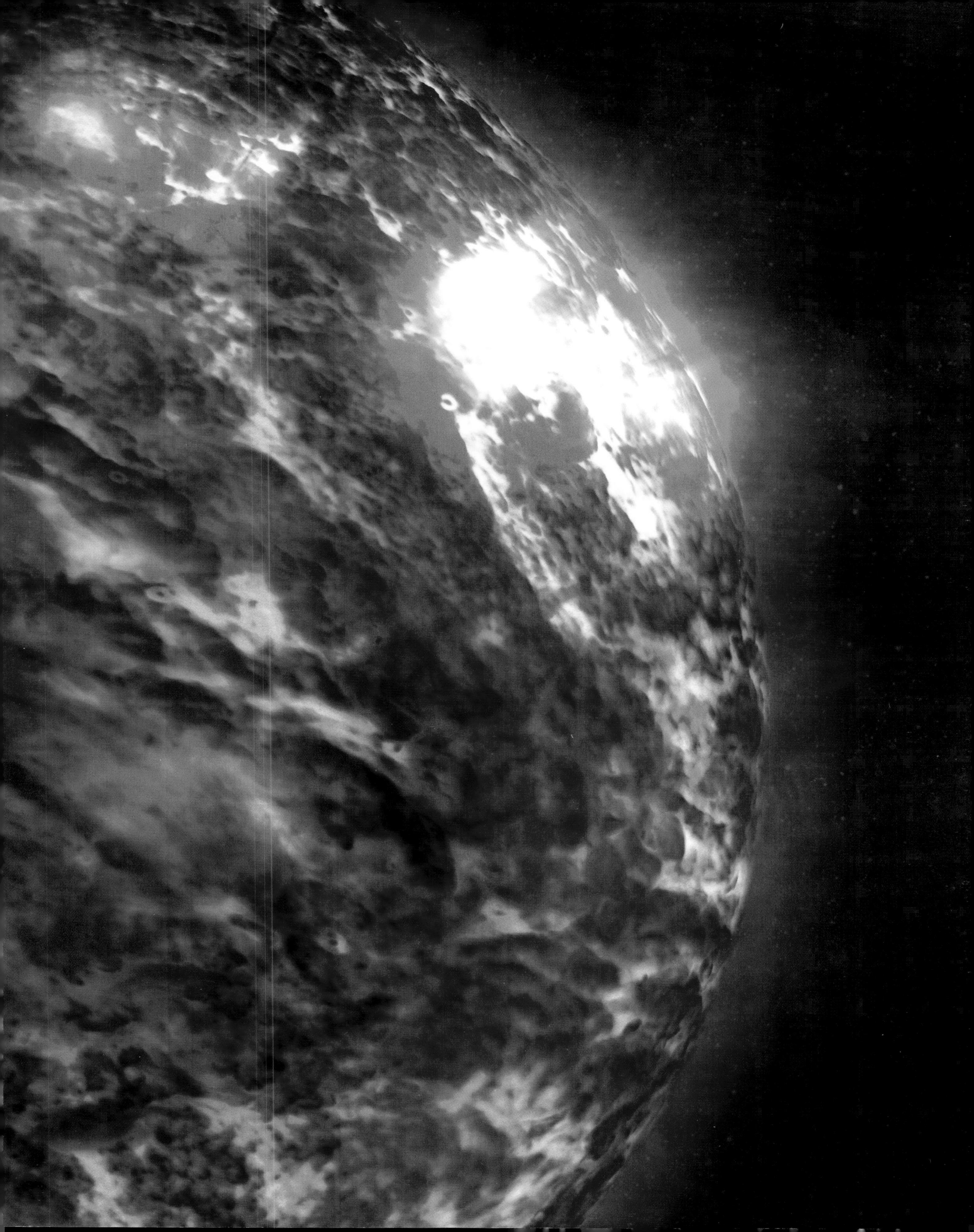

STAR DEATH

When making a television documentary, you are always looking for visual ways to tell complex stories. While filming *Wonders of the Universe*, we journeyed all over the world in search of analogies and backdrops, but for me the most successful of all was an abandoned prison in the heart of Rio de Janeiro, Brazil.

The building itself was a gutted husk, a brick skeleton; all the windows, if it ever had them, were gone. The cells were dormitories of twenty or thirty concrete bunk beds in close rows. Each had a single tiny bathroom, some with ragged pieces of cloth still draped across the entrance, paying lip service to privacy. The walls of the cells were a grotesque patchwork of ripped colour, papered with glamour girls mixed with the odd football team. I found it disturbing for two reasons. First, you can't stop wondering about incarceration there; the centre of a hot, humid city like Rio is not the place to spend years inside a steel and concrete cage. The second was less cerebral: the prison was wired with live explosives. From inside the shell the bright outside pressed and glowed like a stellar surface, impossible to view against the internal black. The light won't come in. It stays outside in the city. I could feel the analogy as I descended down holed, cement-dusted precarious stairwells into the dense heart of the dying star. It is here, inside a violent, condemned structure, far from the light of the surface, that the elements of living things are meticulously assembled. In here, the star transforms from matter consumer to matter producer.

Stars exist in an uneasy equilibrium. Their gravity acts to compress them, which heats them up until the electromagnetic repulsion between the hydrogen atoms is overcome and they fuse together to make helium. This releases energy, which keeps the star up. When the hydrogen runs out, the outward pressure disappears; gravity regains the upper hand and the structure of the star changes dramatically. The core collapses rapidly, leaving a shell of hydrogen and helium behind. Within the shrinking core the temperature rises until, at 100 million degrees Celsius, a new fusion process is triggered. At these temperatures helium nuclei can overcome their mutual electromagnetic repulsion and wander close enough together to fuse – the star begins to burn helium. This transfer from hydrogen to helium fusion has two profound effects: firstly, sufficient energy is

LEFT: Just as in a dying star, the structure of a building and the elements that keep it standing become unstable over time. This prison was given a helping hand to its destruction, but a dying star will detonate itself as it reaches the end of its life, producing spectacular planetary nebulae. It took seconds to demolish the prison block, which is the same length of time it takes for a red giant star to collapse.

released to halt the stellar collapse, so the star stabilises and rapidly swells. This is the beginning of its life as a red giant. Secondly, it fuses into existence the element vital for life. At first sight the fusion of two helium nuclei, each consisting of two protons and two neutrons, should only be able to produce the isotope beryllium-8, composed of four protons and four neutrons. This is an unstable isotope of beryllium that quickly breaks down, but in the intense temperatures of a dying star, as the core exceeds 100 million Kelvin, these nuclei live just long enough to fuse with a third helium nucleus, creating the precious element carbon-12. This is where all the carbon in the Universe comes from; every carbon atom in every living thing on the planet was produced in the heart of a dying star.

The helium-burning phase doesn't end with the alchemic synthesis of carbon, because during the same intensely hot phase in the star's life the conditions allow a nucleus of helium to latch onto a newly minted carbon nucleus to create another element vital for life. Oxygen makes up 21 per cent of the air we breathe, is a prerequisite for water, the solvent of life, and is the third-most common element in the Universe after hydrogen and helium. As you breathe in around two and a half grams of oxygen each minute, it's worth remembering that all this life-giving gas was created in an environment as far away from our understanding of what is habitable as you can get.

Compared with the lifetime of a star, this stellar production line of carbon and oxygen is over in the blink of an eye. Within about a million years the helium supply in the core is used up, and for many stars that's where fusion stops. Any average-sized star, like our sun, has by now reached the end of its productive life. When our sun reaches this stage, in about ten billion years' time, there won't be enough gravitational energy to compress the core any further and restart fusion. Instead, the star becomes more and more unstable, huge pressure points will build up, until eventually the whole stellar atmosphere explodes, hurling the precious cargo of oxygen, carbon, hydrogen, and all, on its journey into space. For at this brief moment in time, no more than a few tens of thousands of years, a dying star will create one of the most beautiful structures in our universe: a planetary nebula.

Once this brief cosmic light show is over, an average-sized star will shrink to an object no bigger than Earth. A white dwarf is the fate of such stars and billions like it, but for massive stars like Betelgeuse the action is far from over. If a star has a mass half as big again as our Sun, it will continue down the chemical production line. As helium fusion slowly comes to an end, gravity takes over and the collapse of the core restarts. The temperature rises, launching the third stage in the birth of our universe's elements, and with temperatures reaching hundreds of millions of Kelvin, carbon fuses with helium to make neon, neon fuses with more helium to make magnesium, and two carbon atoms fuse to make sodium. With more and more elemental ingredients entering the cooking pot, and temperatures rising, the heavier elements are produced one after another. The core continues to collapse, the temperature continues to rise, and the next stage of fusion begins, leaving layers of newly minted elements behind.

With the first twenty-five elements now created within the star, the runaway production line hits a block at the twenty-sixth element, iron, created from a complex cascade of fusion reactions fuelled by silicon. At this stage the temperature of the star is at least 2.5 billion Kelvin, but it has nowhere else to go. The peak of nuclear stability has been reached, and no more energy can be released by adding more protons or neutrons to iron. The final stage of iron production lasts only a couple of days, transforming the heart of the star into almost pure iron in a desperate bid to release every last gasp of nuclear binding energy and stave off gravity. This is where the fusion process stops; once the star's core has been fused into iron, it has only seconds left to live. Gravity must now win, and the star collapses under its own weight forming a planetary nebula.

As I walked away from the prison for the cameras, a button was pressed and the building fell. The demolition took seconds – the same time it takes a red giant star like Betelgeuse to collapse ◉

PLANETARY NEBULAE

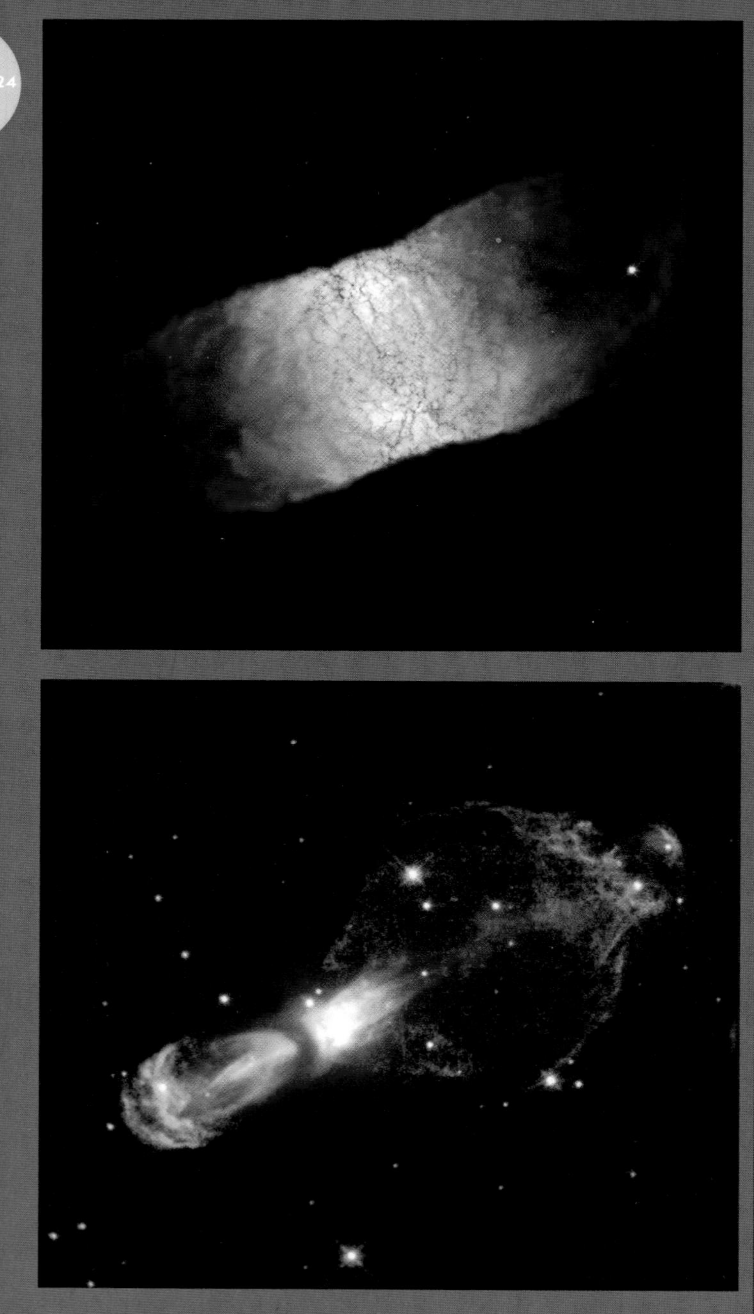

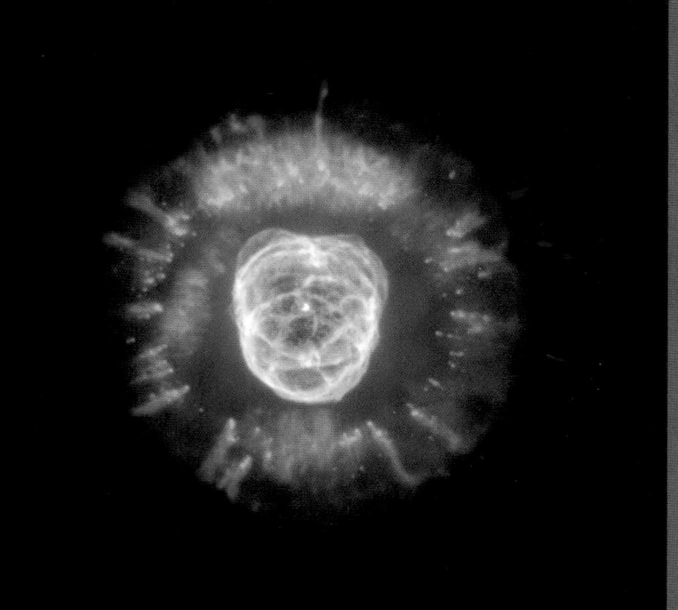

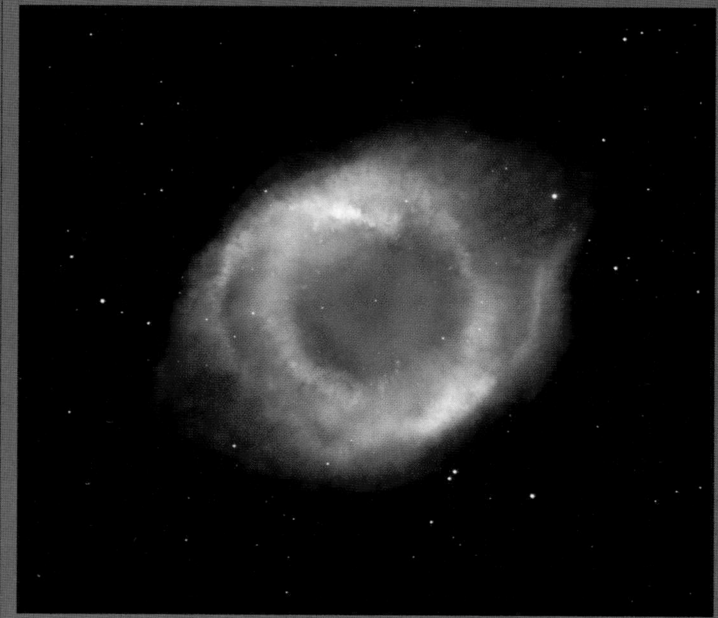

TOP:
This dying star, IC 4406, like many planetary nebulae, is highly symmetrical. It is known as the 'Retina Nebula' because the tendrils of dust emitted from it that have been compared to the eye's retina.

BOTTOM:
About 5,000 light years (4,700 trillion kilometres/2,900 trillion miles) from Earth lies the Calabash Nebula. This image, captured by the Hubble Space Telescope, shows material being ejected from the star.

TOP:
The Eskimo Nebula is so-called because of its resemblance to a head surrounded by fur-lined hood when viewed from Earth. It was discovered in 1787 by astronomer William Herschel.

BOTTOM:
This composite image depicts the Helix Nebula. This planetary nebula resembles a doughnut, as seen from Earth, but new evidence suggests that the Helix in fact consists of two gaseous discs.

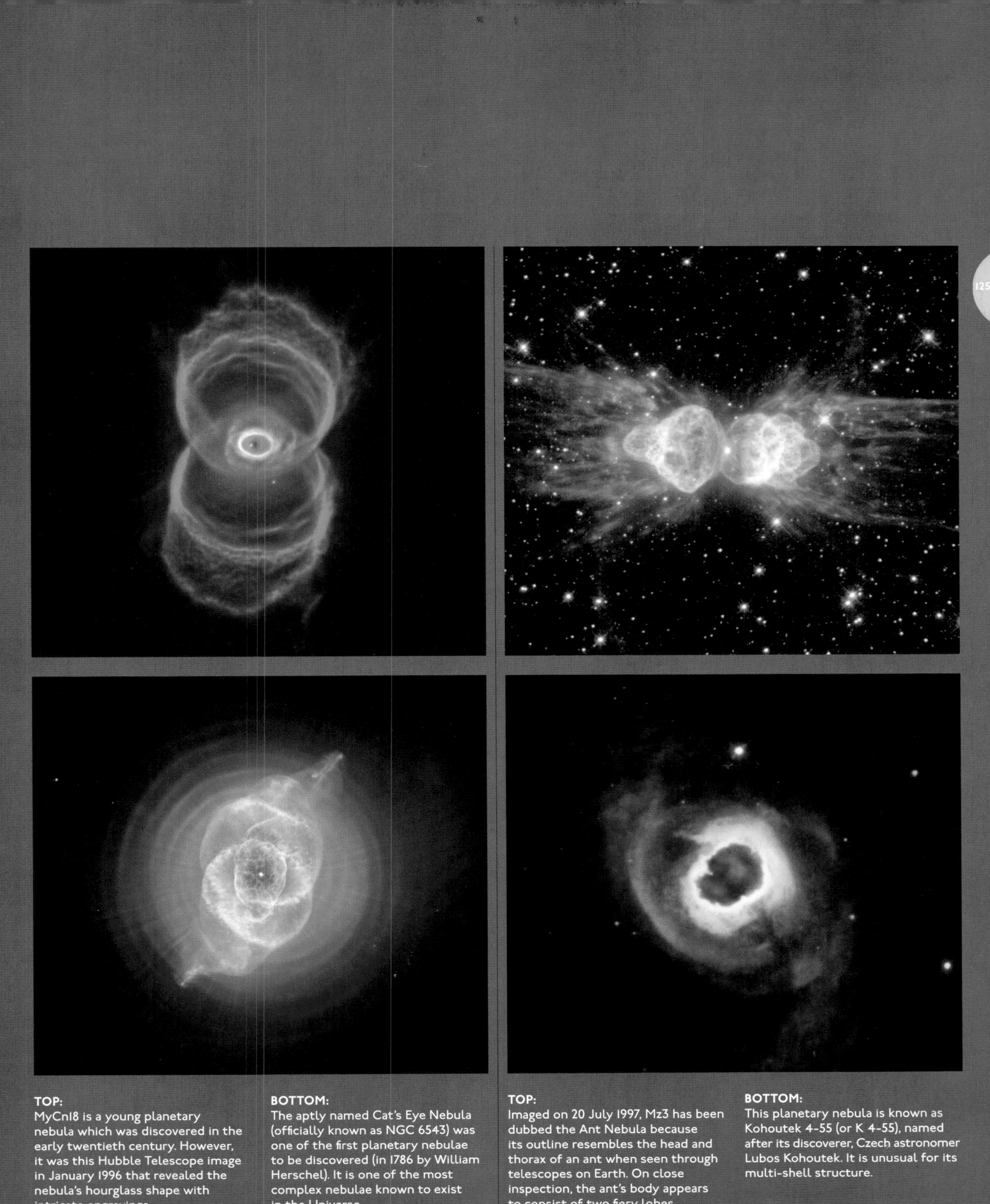

TOP:
MyCn18 is a young planetary nebula which was discovered in the early twentieth century. However, it was this Hubble Telescope image in January 1996 that revealed the nebula's hourglass shape with intricate engravings.

BOTTOM:
The aptly named Cat's Eye Nebula (officially known as NGC 6543) was one of the first planetary nebulae to be discovered (in 1786 by William Herschel). It is one of the most complex nebulae known to exist in the Universe.

TOP:
Imaged on 20 July 1997, Mz3 has been dubbed the Ant Nebula because its outline resembles the head and thorax of an ant when seen through telescopes on Earth. On close inspection, the ant's body appears to consist of two fiery lobes.

BOTTOM:
This planetary nebula is known as Kohoutek 4-55 (or K 4-55), named after its discoverer, Czech astronomer Lubos Kohoutek. It is unusual for its multi-shell structure.

BELOW AND RIGHT: Once the centre of the great American gold rush, the 16-1 mine is one of the few gold mines still operating in the state of California today. Digging for gold there with the miners was an enlightening experience. As I peered at seemingly ordinary rocks, I could see glints and glimmers of a familiar yellow colouring, revealing the stones' precious hidden cargo of gold.

The first twenty-six of the elements are forged in the cores of stars and are distributed through the Universe in their inevitable collapse. But what of the other seventy-two – some of which are vital for life, and many of which we hold most precious? If they are not formed within stellar furnaces, what could their origin possibly be?

In the remote forests of northwestern California, the mountains still hide a secret that made the quiet pine woods the ultimate destination for fortune seekers only a century ago. Although they're empty today, in the late nineteenth century this was the centre of the California gold rush. Hundreds of thousands of people arrived here, trying anything and everything to get rich, from simple panning to the most advanced mining techniques available. Gold worth billions of dollars was extracted, fuelling the rise of one of the world's great cities, San Francisco. The insatiable appetite for gold has waned today, but in the forests around Lake Tahoe, the 16–1 mine remains one of the few gold mines still operating in the state of California.

For almost 100 years, miners have been digging for gold in the 16–1, and it is still one of the richest gold deposits in the world, due to a quirk in the local geology. The unique thing about California is that it sits on the divide between the North American tectonic plate and the Pacific tectonic plate. The

whole region is one enormous fault line, with thousands of smaller faults running through the rocks of the mountains. When you travel into the mine, which is nothing more than a series of horizontal tunnels at gentle gradients hollowed out of the mountainside, you can see these fault lines everywhere; they reveal their presence as visible boundaries between rock and quartz – a maze of mini-faults. One hundred and forty million years ago, in the Jurassic period when the dinosaurs were running around above the mine, hot water bubbled up and flowed through this rock, carrying a precious cargo. Its water was laden with gold brought up from deep within the Earth, deposited through the seams of quartz. For the last 100 years all the miners have had to do is to follow quartz seams laced with shimmering gold.

The gold that runs all the way through the quartz in the 16–1 mine is unusually pure, at anything up to 85 per cent, and the thick tendrils snaking through the rock glint and glimmer that familiar yellow in the sunlight. The rest is about 14.5 per cent silver, with traces of heavier metals. The area is so rich in gold that it can even be found as simple pure nuggets that can be picked up off river beds, and at the 2010 price of around £900 per troy ounce, it's obvious why mines like this are still in operation.

If you stop to think about it though, there's something a bit odd about the value we attach to gold. Throughout history

All the gold dug out of the ground throughout all of human history would just about fill three Olympic-sized swimming pools. It is this almost vanishing scarcity that makes gold so valuable.

people have gone to extraordinary lengths to get their hands on it, which is odd because it isn't particularly useful for anything. Copper and iron will help you survive, but gold is next to useless. Most of the gold that we've struggled to extract has ended up as jewellery. The only thing that gold has going for it, other than being shiny, is that it is incredibly rare, and this is what drives up its price. All the gold dug out of the ground throughout all of human history – with all the associated tragedy and elation, hardship and riches – would just about fill three Olympic-sized swimming pools.

It is this almost vanishing scarcity (three swimming pools relative to the size of a planet) that makes gold so valuable; it is just one of many rare elements that are to be found in the most minute of traces within the Earth.

There are over sixty elements heavier than iron in the Universe, some are valuable, such as gold, silver and platinum; some are vital for life, such as copper and zinc; and some are just useful, such as uranium, tin and lead. Very massive stars can produce very tiny amounts of the heavier elements up to bismuth-209 (element number 89) in their cores by a process called neutron capture, but it is known that this makes nowhere near enough to account for the abundances we observe today. There simply haven't been enough massive stars in the Universe.

The conditions necessary to produce large amounts of the elements beyond iron are only found in the most rare of all celestial events. Blink and you'll miss them, because in a galaxy of 100 billion stars the conditions violent enough to form substantial amounts of these elements will exist on average for less than two minutes in every century ◉

SUPERNOVA: LIFE CYCLE OF A STAR

All stars are born from clouds of gas, but the length of their life and their eventual fate are governed by their mass (i.e. how much gas they contain). Stars dozens of times heavier than the Sun live for only a few million years before swelling into supergiants and exploding as supernovae (top row). However, stars like the Sun live longer and die more gently, shining steadily for billions of years before swelling into red giants and losing their outer layers as a planetary nebula (middle row). The core of the star, exposed as a white dwarf, then continues to glow for billions of years more before gradually fading out. The least massive stars, the red dwarfs (bottom), simply fade out over tens of billions of years.

HIGH-MASS STAR RED SUPERGIANT

Supergiant becomes mor red as it starts to cool

1,000 MILLION KM

NEBULA

Dense nebula area begins to contract

200,000 BILLION KM

PROTOSTAR

Gravitational tug will cause it to collapse and central temperature to rise to around 15 million °F

100 MILLION KM

Star emits heat and light caused by nuclear fusion

1 MILLION KM

SUN-LIKE STAR

Star becomes a red giant as hydrogen-shell burning begins

LOW-MASS STAR RED DWARF

Shines for trillions of years, never becoming a red giant

100,000 KM

Star explodes as a supernova and its outer layers are blown off, producing elements heavier than iron

If star's mass is over 1.4 solar masses, it will collapse and become a neutron star. These are very dense and compact

15 KM

If the remnant is above 3 solar masses it will collapse and become a black hole. These are regions of space in which gravity is so strong that not even light can escape, surrounded by swirling discs of captured gas and dust

50 KM

Red giant outer layers start to form planetary nebula

Star collapses after burning its helium shell to become a white dwarf

10,000 BILLION KM

This will eventually fade to become a black dwarf

10,000 KM

Star starts to collapse as hydrogen is used up

10,000 KM

Star becomes extremely dense with a faint core. This will eventually turn into a white dwarf

Core stops glowing

THE BEGINNING AND THE END

BELOW AND RIGHT: This computer-generated sequence of images shows what will happen when Betelgeuse goes supernova. Deep in the heart of the star, the core will succumb to gravity and fall in on itself, then rebound with colossal force. The blast wave emitted generates the highest temperatures in the Universe. Over millions of years the scattered elements of the exploded star will become a nebula, at the heart of which is a super-dense core that is Betelgeuse the neutron star.

After a few million years of life, the destiny of the largest stars in our universe is a dramatic one. Having run out of hydrogen and burnt through the elements all the way to iron, giant stars teeter on the edge of collapse. Yet even in this dilapidated state these stars have one last violent act, and it is a generous one. It occurs with such intensity that it allows for the creation of the heavy elements.

If we could gaze deep into the heart of one of these dying giants, we would see the core finally succumb to gravity. As fusion grinds to a halt, this giant ball of iron falls in on itself with enormous speed, contracting at up to a quarter of the speed of light. This dramatic collapse causes a rapid increase in temperature and density as the core shrinks to a fraction of its original size. The inner core may eventually shrink to 30 kilometres (19 miles) in diameter. At this point, with temperatures nearing 100 billion Kelvin and densities comparable to those inside an atomic nucleus, quantum mechanics steps in to abruptly halt the collapse. By now most of the electrons and protons in the core have been literally forced to merge together into neutrons. Neutrons, in common with protons and electrons, obey something called the Pauli exclusion principle, which effectively prevents them from getting too close to one another (in more technical terms, no two neutrons can be in the same quantum state). This has the effect of making a ball of neutrons the most rigid material in the Universe – 100 million million million times as hard as a diamond. When the neutrons can be compressed no more, the contraction must stop and all the superheated collapsing matter rebounds with colossal force. A shockwave shoots out through the star and as this blast wave runs into the outer layers of the star it generates the highest temperatures in the Universe – 100 billion degrees. The precise mechanism for

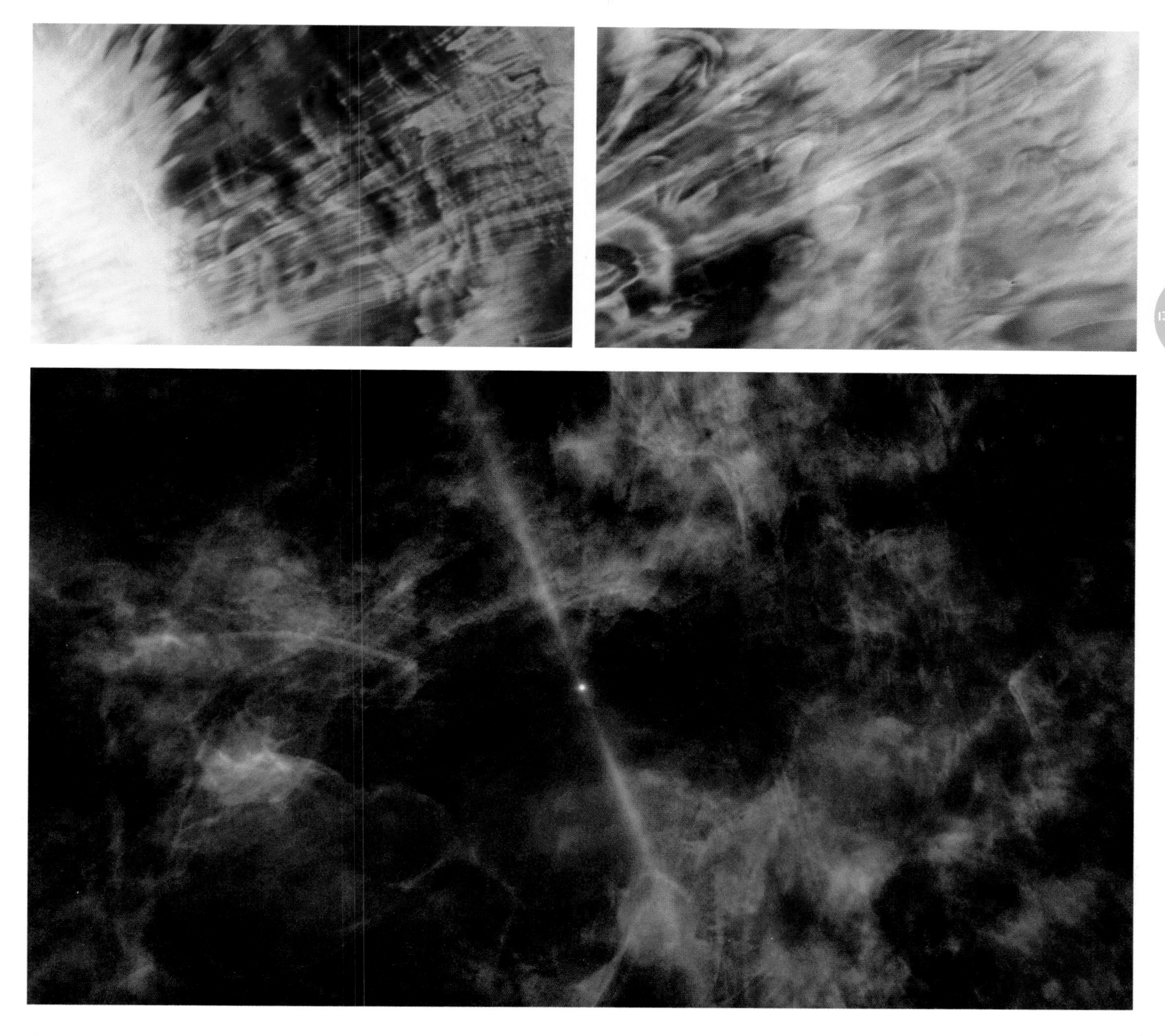

this rapid heating is not fully understood, but it is known that for a matter of seconds the conditions are intense enough to form all the heaviest elements we see in our universe, from gold to plutonium. This is a Type II supernova – the most powerful explosion we know of.

Supernovae are so rare that since the birth of modern science we have never had the chance to see one close up. The last supernova explosion seen from Earth in our galaxy was in 1604, a few years before the invention of the astronomical telescope. On average, it is expected there should be around one supernova explosion in the Milky Way per century, but for the last 400 years we've had no luck. It's long overdue and astronomers are always searching the skies for stars which they think might be the most likely candidate to go supernova.

One of the prime candidates is Orion's shining red jewel, Betelgeuse. With so many telescopes trained on this nearby star, we have been able to follow its every move for decades. Charting its brightness, we have discovered that it is extremely unstable; it has dimmed by about 15 per cent in the past decade. As supernova candidates go, Betelgeuse is top of the list. It is generally thought that Betelgeuse could go supernova at any time. It is a relatively young star, perhaps only ten million years old, and has sped through its life cycle so rapidly because it is so massive. However, when you're ten million years old, the end of your life can be quite drawn out and a phrase like 'any time soon' in stellar terms is not quite what you might expect. It means that Betelgeuse should go supernova at some point in the next million years, but equally it could explode tomorrow. What we do know is that when it does go it will provide us with quite a show. Betelgeuse is only 500 light years away, almost uncomfortably close, which means that the explosion will be incredibly bright. It will be

LEFT: The giant Orion Molecular Cloud is an extensive area of star formation about 1,500 light years from us, centred on the impressive Orion Nebula. This infrared image of it, from NASA's Spitzer Space Telescope, shows light from newborn stars within the Orion Nebula. The nebula can be seen from Earth with the naked eye as a hazy 'star' in Orion's sword.

BELOW: This computer-generated image shows just how bright scientists believe the heavens will be once Betelguese has gone supernova; it will flood the skies with light – day and night.

BOTTOM: When stars are more massive than about eight times the Sun, they end their lives in a spectacular explosion. The outer layers of the star are hurtled out into space at thousands of miles an hour, leaving a debris field of gas and dust. Where the star once was, a small, dense object called a neutron star is often found. While around only 16 kilometres (10 miles) across, the tightly packed neutrons it contains have more mass than the entire Sun. The bright blue dot in the centre of this X-ray image of RCW 103 is believed to show the neutron star that formed when the star exploded in a supernova 2,000 years ago.

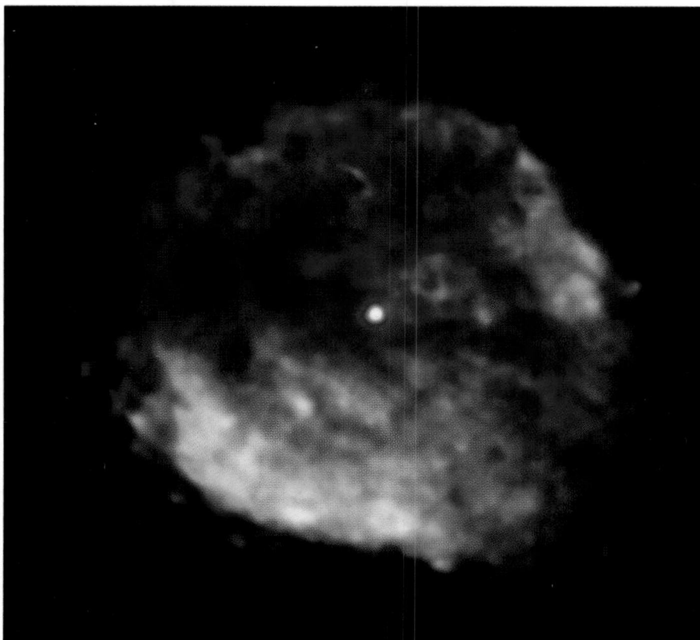

by far the brightest star in the sky and it may even shine as brightly as a full moon at night and fill the sky as a second sun during the day.

In a single instant, Betelgeuse will release more energy than our sun will produce in its entire lifetime. As the explosion tears the star apart, it will fling out into space all the elements the star has created through its life.

Over millions of years these newly minted elements will spread out to become a nebula, a rich chemical cloud drifting in space. At the heart of it, all that will remain will be the super-dense core of neutrons; the remnants of the star that was once a billion miles across will have been squashed out of all recognition by gravity. This is a neutron star, the ultimate destiny of Betelgeuse; a dense, hot ball of matter which is the same mass as our Sun but only 30 kilometres (19 miles) across.

We may not have seen neutron stars close up, but we have seen them from afar. X-ray images have been taken that give us vital information about these stars, in particular recent pictures of RCW 103, the two-thousand-year-old remnant of

When Betelgeuse explodes it will be incredibly bright. It will be by far the brightest star in the sky and it may even shine as brightly as a full moon at night and fill the sky as a second sun during the day.

a supernova explosion that occurred about 10,000 light years from Earth (see left).

This may sound like a cosmic graveyard, but it is in the deaths of old stars that new stars are born. This is the Earthly cycle of death and rebirth played out on a cosmic scale. We can see that beautiful cycle happening today in the constellation of Orion. In an area known as the sword handle lies the Orion Nebula. To the naked eye it appears to be a misty patch of light in the night sky, but through a telescope it is a majestic wonder of the Universe. Hidden in its clouds are bright points of light, new stars forming from the clouds of elements blown out by supernova explosions; the new born from the deaths of the old.

It is from such a cycle that we emerged – within a nebula just like this, five billion years ago, our sun was formed. Around that star a network of planets condensed from the ashes, and amongst them was Earth; a planet whose ingredients originated from the nebula, a cloud of elements formed in the deaths of stars, drifting through space.

But that's not quite the end of the story, because it is now thought that the chemical elements themselves are not the most complex pieces of 'us' that were assembled in the depths of space ◉

THE ORIGIN OF LIFE

At first sight the graph opposite – depicting the spectrum of the light from the Orion Nebula (taken from the Herschel Space Observatory Telescope) – looks rather uninspiring, but the information that it contains is in fact fascinating. This illustration reveals that the Orion Nebula is not just a cloud of elements; there is complex chemistry happening out there deep in space.

Just like the black lines in the spectrum of the Sun, the peaks on this graph correspond to particular chemical elements, but some of these peaks derive from complex molecules – there is water in the nebula, and sulphur dioxide. Perhaps more surprisingly, there are also complex carbon compounds – methanol, hydrogen cyanide, formaldehyde and dimethyl ether. This is direct evidence for complex carbon chemistry occurring in deep space. This is tremendously exciting because it means that we are seeing the beginnings of the chemistry of life in a vast cloud of interstellar gas.

The connection doesn't end there; we may be connected to the chemistry out there in space even more directly. The photo opposite is of a meteorite, a piece of rock that fell to

BELOW: There are thousands of asteroids in our solar system, mostly within an asteroid belt that formed 4,568 million years ago, and on average one meteorite falls to Earth once a month. However, each and every one discovered is hugely important, regardless of its size, as these asteroid pieces give us a real insight into what forms the building blocks of life.

RIGHT: This seemingly ordinary piece of rock is anything but; this asteroid fragment is older than any rock on Earth and is one of the thousands of meteorites that fall onto our planet every year.

The fundamental building blocks of life may have formed in the depths of space and been delivered to our planet by meteorites.

Earth from somewhere out in the depths of the Solar System. It is almost certainly older than any rock on Earth because it was formed from the primordial dust cloud, the nebula that collapsed to form the Sun and the planets five billion years ago. When looking inside this ancient rock we discovered something incredibly interesting: it was found to contain amino acids, the building blocks of proteins, which in turn are the building blocks of life. This strongly suggests there was very complex carbon chemistry happening out there in space, forming the building blocks of life, over four and a half billion years ago. It raises the intriguing prospect that the first amino acids on Earth may have formed in the depths of space and been delivered to our planet by meteorites.

This is one more beautiful piece of evidence that forces us to think differently about those twinkling lights and smudges of gas and dust in the sky. When we look out into space we are looking at our place of birth. We truly are children of the stars, and written into every atom and molecule of our bodies is the history of the Universe, from the Big Bang to the present day ◉

ORION'S MOLECULAR MAKE-UP: This detailed spectrum, obtained by ESA's Herschel Space Observatory, shows the fascinating chemical fingerprints of potential life-enabling organic molecules in the Orion Nebula.

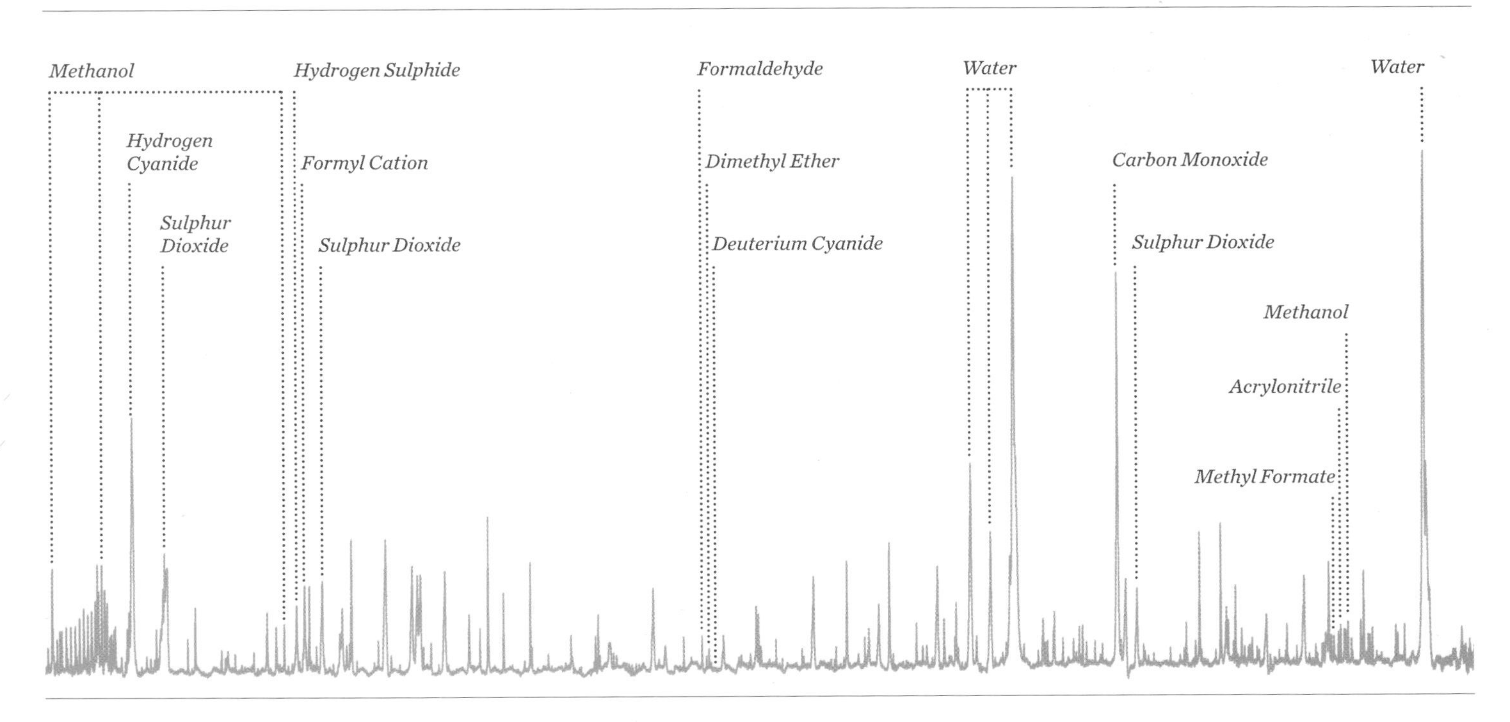

Our story is the story of the Universe. Every piece of every one and every thing you love, of every thing you hate, of every thing you hold precious, was assembled in the first few minutes of the life of the Universe, and transformed in the hearts of stars or created in their fiery deaths. When you die those pieces will be returned to the Universe in the endless cycle of death and rebirth. What a wonderful thing to be a part of that universe – and what a story. What a majestic story!

RIGHT: Supernovae are the long-awaited spectacles of the skies. It is in the death of old stars that new ones are born, and their demise plays a crucial part of the endless cycle of death and rebirth that occurs right across our universe.

CHAPTER 3

FALLING

FULL FORCE

For all its scale and grandeur, the Universe is shaped by the action of just four forces of nature. Two of these, the weak and strong nuclear forces, remain hidden from everyday experience inside the atomic nucleus. The third force, electromagnetism, is perhaps most familiar to us, as it is the one we marshal to power our lives – electric currents flow because of the action of this force. Finally, there is gravity, the great sculptor – the force that acts between the stars. Gravity shapes the cosmos on the largest distance scales. From the formless clouds of hydrogen and helium that once filled our universe, gravity forged the first stars, sculpted the first planets and arranged them into the exquisite shapes of the galaxies. Having assembled countless billions of solar systems, gravity drives their cycles and rhythms. It is the invisible string behind the revolution of every moon around every planet and every planet around every star. Gravity keeps our feet on the ground and the Universe ticking over.

BOTTOM: Before embarking on the voyage of their lives, astronauts are prepared for the flight and the sensation of weightlessness in aircrafts such as this C-131 at Wright Air Development Center, which flies at a 'zero-g' trajectory. These flight simulators are dubbed 'vomit comets' because of the nausea they often induce.

Gravity is more than a mere gentle presence; it is relentless, and for the largest agglomerations of matter in the Universe – the stars – it is both creator and destroyer. Stars shine in temporary resistance to gravitational collapse, but when they run out of nuclear fuel and the other three forces can no longer rearrange the matter in their cores in order to release energy and resist its inward pull, gravity crushes the most massive of them out of existence. In doing so, it creates the least understood objects in the Universe.

Soviet cosmonaut Gherman Titov is perhaps not the luckiest of men. In 1960 he was selected alongside Yuri Gagarin for the Soviet manned space programme. Out of the twenty men who started the programme, only these two made it through a fierce selection process that tested their physical and psychological resilience to the limit. Throughout training the two fighter pilots matched each other point for point, but someone had to be first, and Gagarin was given the ticket into the history books. On 12 April 1961, Gagarin became the first human to travel into space, completing a single orbit in 108 minutes before returning to Earth first in Vostok 1 and then by parachute. In one of those interesting bits of space trivia, Gagarin actually arrived back on Earth after his spacecraft, because he ejected at an altitude of 7,000 metres (23,000 feet) due to worries about the safety of the capsule on landing. Vostok 1 arrived safely on the ground 10 minutes before he did.

BELOW: This bus ride to the Vostok launch on 12 April 1961 was the first part of the journey that was to make Yuri Gagarin a Soviet hero and worldwide celebrity.

BELOW: Astronauts prepare for Extravehicular Activity by practising techniques on a Hubble Space Telescope mock-up in the Neutral Buoyancy Laboratory. Underwater conditions simulate the weightlessness experienced in space.

RIGHT: The race for space was on in the 1960s, as the US and Soviet nations battled to be the first to launch a human being into space.

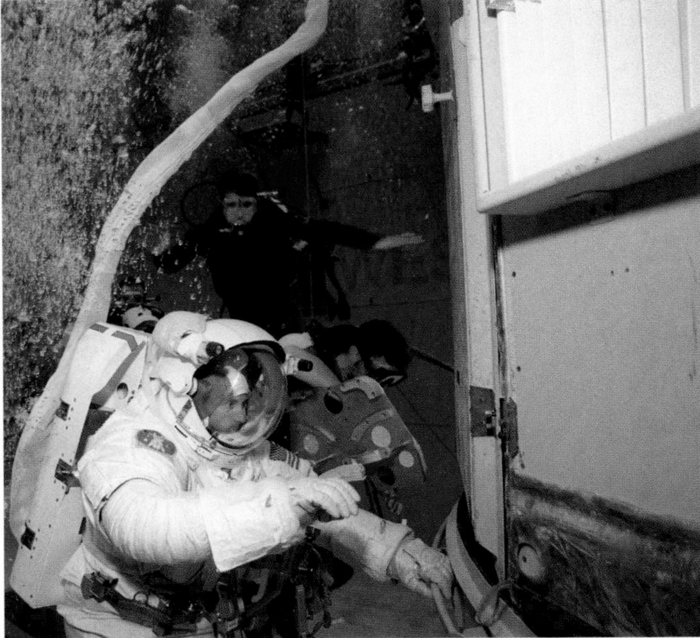

On his return, Gagarin became a Soviet hero and a worldwide celebrity, leaving Titov to become the second man to orbit our planet. Titov's name will be unfamiliar to most, although to this day he remains the youngest man ever to make the journey into space, at just under 26 years old. He piloted Vostok 2 on 6 August 1961, completing 17 orbits of Earth. Titov also claimed a rather less glamorous place in the history books; on the 25.3-hour mission, he not only became the first man to sleep in space (snoozing for a couple of hours as his spacecraft orbited the planet), but also the first to suffer the symptoms of a condition that has affected almost half of those who have experienced weightlessness for an extended period of time. Titov was the first victim of Space Adaption Syndrome. Known more usually as space sickness, this condition includes a variety of symptoms such as nausea, vomiting, vertigo and headaches as a common reaction to the odd sensations of space travel. Although weightlessness remains one of the great thrills of being an astronaut, it is also one of the most difficult to prepare for. Since Titov introduced medics to Space Adaption Syndrome, space agencies around the world have employed the only method they can of creating weightlessness here on Earth. How is it possible to remove the effects of gravity? The answer is by doing the same thing that Gagarin and Titov did: by falling towards Earth.

The American response to the Vostok programme was Project Mercury, a series of six manned launches which included the historic flights of Alan Shephard, the first American in space, on 5 May 1961, and John Glenn, the first American to orbit Earth. The astronauts selected for the programme, known as the 'Mercury Seven', became celebrities in the United States, and all of them eventually flew into space. The final flight of the Mercury Seven was John Glenn's Space Shuttle mission in 1998, which he completed at the age of 77. The Tracy brothers in the TV series *Thunderbirds* were named after five of the Mercury Seven: Scott (Carpenter), Virgil ('Gus' Grissom), Alan (Shephard), Gordon (Cooper) and John (Glenn). Wally Schirra and Deke Slayton missed out. (I think Wally and Deke would have been great names for *Thunderbirds* pilots. The days when astronauts were bigger than rock stars are sadly missed.)

During training for Project Mercury, perhaps after hearing about the experiences of Titov, NASA developed a way of flying a regular military aircraft to take would-be astronauts on an unusual ride. Using a C-131 aircraft, weightlessness was achieved by flying an unconventional flight path. This parabolic path creates a brief period of around 25 seconds during which all the occupants of the plane experience the sensation of weightlessness. This is because they are actually weightless; it may be brief, but when repeated twenty or thirty times in succession, the physiological effects are just as intense as those felt in space. This led to the C-131 being named the 'Vomit Comet', a name that has stuck with every plane used for this task ever since.

I've known about the Vomit Comet since I was a child, because I was, and still am, passionate about the space programme. Imagine my delight when I heard we were going to ride in it for our film on gravity. Who cares if it makes you feel rough, if the Mercury Seven could face it, so could I.

The Vomit Comet is the perfect place to experience the two related aspects of the force of gravity that hold the key to

Feature Index

	Page		Page
Abby	14	Editorials	4
Amusements	6	Sports	22
Comics	25	Society	5
Crossword	18	Want Ads	26
Jumble	14	Radio-TV	26

28 PAGES TODAY

The Huntsville Times

Where Progress... Covers The Valley!

VOL. 51, NO. 21 — CHICAGO DAILY NEWS SERVICE — HUNTSVILLE, ALABAMA, WEDNESDAY, APR. 12, 1961 — ASSOCIATED PRESS — WIREPHOTO — 45c PER WEEK

Man Enters Space

'So Close, Yet So Far,' Sighs Cape

U. S. Had Hoped For Own Launch

CAPE CANAVERAL, Fla. (AP) — The Redstone rocket which the United States had hoped would boost the first man into space stands on a launching pad here. The Soviet Union beat its firing date by at least two weeks.

"So close, yet so far," commented a technician who is helping groom the Redstone to send one of America's astronauts on a short suborbital flight, hopefully late this month or early in May.

"If we hadn't had those troubles last fall and on the chimp and Little Joe shots this year, we might have made it," the technician said.

"But you have to give the Russian scientists credit. They've accomplished a remarkable breakthrough."

Dr. Hugh Dryden, deputy director of the National Aeronautics and Space Administration, told the House Space Committee in Washington Tuesday that the earliest possible date for the manned launching is about April 28.

Project Mercury officials had hoped to achieve a manned Redstone flight last December or January. A series of launch mishaps necessitated additional launchings to qualify the system.

On Nov. 8, a space capsule failed to separate from a Little Joe rocket fired from Wallops Island, Va., in a test of the escape system.

Two weeks later, a Redstone fizzled because of a faulty connection which caused the escape tower to fire, leaving the rocket and capsule on the pad. This test had to be repeated before Ham, the space chimpanzee, was sent up on a short trip Jan. 31.

An engine thrust regulator stuck on the chimp shot, creating excessive thrust which lofted the chimp, Ham, higher and farther than intended. Another Redstone was fired to prove out corrections made in the regulator, again delaying the manned trip.

Another setback occurred March 18 when a repeat of the Little Joe escape test fizzled. Another try is set for about April 20 and must succeed before the Redstone now on the pad hurls aloft an astronaut.

* * *

By ARTHUR J. SNIDER
Chicago Daily News Science Writer

The momentous Russian achievement has taken some of the zest out of the buildup for the first American astronaut's rocket hop down the Atlantic missile range.

There had been hope, swelling each day, that the United States might slip the first human being into space ahead of the Soviets.

For many months — ever since last fall — it had been expected that the Russians would cash in on their capability and launch a

Hobbs Admits 1944 Slaying

By BOB WARD
Of The Times Staff

Isham D. Hobbs confessed today to the brutal murder in 1944 of Mrs. Margaret Thornton Fleming, Circuit Solicitor Macon L. Weaver said.

Hobbs, now 43, is held by Air Force authorities at Eglin Air Force Base, Fla. He signed a statement there detailing his knife-slaying of the prominent 52-year-old widow, Weaver said he learned from Air Force officials.

* * *

The suspect, who has undergone psychiatric treatment by military authorities since Feb. 6, has recovered his memory in full, Weaver was told. Psychiatrists said Hobbs' apparent amnesia resulted from "hysteria" rather than from any medical cause.

Hobbs, who attempted suicide last November at Bartow, Fla., and was then exposed as a long-time fugitive, will be returned to MacDill Air Force Base, near Tampa, Fla., from Eglin AFB.

Weaver plans to travel to MacDill tomorrow, he said.

Hobbs, accused also of deserting from the Army Air Corps in October, 1943, reportedly will be court-martialed, and released to civil authorities here. He was wanted for desertion and was still at large when the murder charge was brought against him in May 1944.

* * *

Hobbs told Eglin authorities he was living in a cave in the mountainous region near Mrs. Fleming's home north of Farley when the killing occurred, Weaver said.

He stated he went to the Fleming home in an effort to get a shotgun he believed to be there. Finding Mrs. Fleming's daughter, Vivian, asleep when he broke into the house before dawn, he decided to knock her unconscious and spirit her away to his cave, he stated.

The blow he struck the 21-year-old girl [illegible] only stunned her, and her screams attracted Mrs. Fleming [illegible]

This is Russian Maj. Yuri Gagarin, history's first man in space. The Russians today rocketed him around the earth in an orbit taking slightly less than 90 minutes and brought him back safely to a prearranged spot in the Soviet Union. (AP Wirephoto via radio from Moscow)

Praise Is Heaped On Major Gagarin

'Worker' Stands By Story

LONDON (AP) — The Daily Worker, Communist party paper in Britain, said today it is standing by its story that the Soviet Union launched a man into space last Friday.

* * *

A spokesman for the editor said: "Our story came from good sources. All we know is what we published today. Now of course there is this one."

By "this one," he referred to the Moscow announcement that Maj. Yuri A. Gagarin made an orbital flight around the earth today in a five-ton space ship and returned safely.

* * *

The Worker said in this morning's edition that the Soviet Union sent an astronaut aloft last Friday on three trips round the world, brought him back alive and then put him under treatment for aftereffects. He was identified in the story as the test pilot son of a top-ranking Soviet air

First Man To Enter Space Is 27, Married, Father Of Two

LONDON (AP)—Moscow television presented a picture of the Soviet Union's first space man today, describing him as a man with "a good honest smile."

The portrait of Maj. Yuri A. Gagarin was shown and then came this broadcast comment, repeated by Moscow radio:

"For those who did not see this picture we should like to give a description of this splendid man.

"On the screen appears the image of a man aged about 25-28 with a kind, Russian face, eyes set well apart, fine bushy brows and high forehead.

"He wears a flying helmet, a light overall suit. He smiles a good, honest smile. And is there any need to add that this man who has been the first to dare to fly to space, to reach for the stars, to look down on our earth, is a man of a very great and very real character. This is evident in his smile, in the intelligent, fine eyes."

* * *

Gagarin was 27 just a month ago.

He is married to Valentina Gagarina, 26, who also has a scientific background. She was graduated from medical school at Orenburg.

They have two daughters, Yelena, 2, and Galya, just a month old.

* * *

The cosmonaut has an ideal so

Reds Deny Spacemen Have Died

By THE ASSOCIATED PRESS

Have some Soviet astronauts been killed in space flight experiments before Yuri A. Gagarin's sensational trip?

No, Soviet officials insist.

But some Western sources say they believe one or a few Russians did perish in unsuccessful attempts. Brig. Gen. Don Flickinger, head of the medical section of the U.S. Air Force astronaut selection and

Soviet Officer Orbits Globe In 5-Ton Ship

Maximum Height Reached Reported As 188 Miles

MOSCOW (AP)—A Soviet astronaut has orbited the globe for more than an hour and returned safely to receive the plaudits of scientists and political leaders alike. Soviet announcement of the feat brought praise from President Kennedy and U. S. space experts left behind in the contest to put the first man into successful space flight.

By the Soviet account, Maj. Yuri Alekseyvich Gargarin, rode a five-ton spaceship once around the earth in an orbit taking an hour and 20 minutes. He was in the air a total of an hour and 48 minutes.

The whole sequence of events and the announcements relating to it raised a number of questions. The Soviet announcement said the flight took place today between 9:07 and 10:55 a.m., but some persons in Moscow's Western colony were skeptical that the feat actually came off today.

There was a curious sequence of events leading up to the announcement.

Rumors had been circulating several days that the space coup had been pulled off. Two days ago, Soviet TV technicians moved into the Central Telegraph Office with the evident purpose of getting pictures of correspondents in action as they reported such a story. There were various reports, none verifiable from official sources, that the flight had been made.

* * *

Then Tuesday night the Daily Worker, London Communist newspaper with apparently sound connections in Moscow, reported that the flight took place last Friday. In splash headlines, the Daily Worker heralded "the first man in space," saying he had completed three orbits before returning to earth suffering from "aftereffects of the flight."

* * *

That led up to today.

About 9:30 a.m., Western correspondents were tipped off to be listening to their radios at 10 a.m. The announcement came at 10 a.m., saying the astronaut still was in orbit. At two intervals the radio broadcast messages, reportedly from him over South America and Africa.

Then came the announcement that the spaceship had been called back to earth.

Some in the Western colony expressed wonderment that the Soviet Union, with its tight control over communications, would take such a chance—announcing the flight before a successful completion.

* * *

As these skeptics saw it, the event would have turned into perhaps the most publicized disaster in history if anything had gone wrong between 10 a.m. and the announced time of landing, 55 minutes later.

VON BRAUN'S REACTION:

'To Keep Up, U. S. A. Must Run Like Hell'

Timesfoto by William McCormick

WERNHER VON BRAUN
He Praises A Russian Achievement

By BILL AUSTIN
Of The Times Staff

A disappointed Dr. Wernher von Braun, arriving in Huntsville today, called Russia's space flight a tremendous thing and labeled it the "shot heard around the world."

"I'm disappointed because here again we came in in second place," he declared.

Von Braun arrived at the Huntsville airport from Grove City, Pa., where he had addressed a college group yesterday.

He said we had hoped all along the United States would be able to place an astronaut up first, but he said Russia has an excellent space program and they demonstrated it by this flight.

"We are going to have to run like hell to catch up," he asserted.

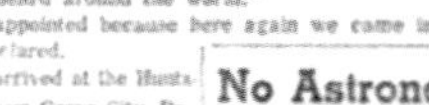

No Astronaut Signal Received At Ft. Monmouth

FT. MONMOUTH, N.J. (AP)—The Astro Observation Center did not receive any radio signals from the Soviet satellite containing the first space navigator, a

understanding what gravity actually is. Firstly, it is possible to completely cancel out the effects of gravity by simply falling towards the ground. This sets gravity apart from all the other forces of nature; it is not possible to negate the effect of electric charge, other than by adding more electric charge of the opposite sign. The Comet achieves the removal of gravity simply by flying along the trajectory that a cannon ball would take when fired out of a gun. The plane doesn't just drop to the ground like a lift with a severed cable, of course (because then it would be impossible to control), but the acceleration of the plane towards the ground is exactly the same as the acceleration you would experience in a falling lift or a parachute jump (if you neglect air resistance). In numbers, the plane must accelerate towards the ground at 9.81 metres per second squared to cancel out the force of gravity. In order to keep the plane under control, it also flies forward at its usual flight speed. This results in the plane flying along a parabolic path. The fact that the effects of gravity are completely removed in freefall is very interesting, and the converse is also true: it is also possible to add to Earth's gravitational pull by accelerating.

Everyone knows that astronauts in space are weightless and float around inside their spacecraft, but not everybody knows why. It is not because they are a long way from Earth that gravity is absent (they are in fact only a few hundred miles above Earth's surface, and the strength of Earth's gravitational field in near-Earth orbit is not too different to the strength on the surface), it is that the effects of gravity are removed by falling, which is important point number one.

We flew in a modified Boeing 727-200, which is still used today for training shuttle astronauts. During the flight I was also able to demonstrate another strange but equally important and related aspect of gravity. Isaac Newton knew it when he wrote down his theory of gravity in 1687, as did Galileo many decades before him. The strange thing is this: all objects fall at the same rate under the force of gravity, even

As we accelerate away from Earth we experience a gravitational pull 1.8 times the strength of Earth – so I weigh nearly twice as much as I do back down on the ground.

though gravity acts on objects in proportion to their mass. Newton and Galileo knew this to be the case because they did experiments and noticed that it was true, but they had absolutely no idea why. If you think about it for a moment, it is very odd indeed. Newton found that the gravitational force between two objects, such as Earth and you, is proportional to the product of their masses. So the force you feel due to the pull of Earth's gravity is proportional to the mass of Earth multiplied by the mass of you. If you were to double your mass, the force between you and Earth would double. But, the rate at which you accelerate towards Earth because of its gravitational pull is also proportional to your mass, and when you work everything out it turns out that your mass completely cancels out, so therefore all things fall at the same rate under gravity. This looks very strange and was famously demonstrated by Apollo 15 Commander Dave Scott on the surface of the Moon in 1971. Scott dropped a feather and a hammer to the ground and, of course, both hit the ground at the same time. The reason you can't do this on Earth is because air resistance slows the feather down, but in the high vacuum of the Lunar surface the only force acting on the falling objects is gravity. No matter how much physics you know, this is entertaining to watch because it isn't in accord with common sense! Surely a cannon ball should fall to the ground faster than a single atom? The answer is, no, it doesn't, and here is something to think about for later on: even a beam of light falls to the ground at the same rate as a cannon ball. Understanding this concept is key to understanding gravity.

I was able to demonstrate this for myself in the Vomit Comet armed with a model of Einstein. When we were weightless, I let a little plastic Albert float beside my head. One way of understanding why we floated next to each other is to simply state that we were both weightless, so we floated, but think about what this looks like from outside the plane. To someone on the ground looking up at us, the plane, myself and plastic Albert are all falling towards the ground under the action of Earth's gravity, and obviously we are falling at the same rate. If I fell faster than Einstein, he wouldn't float next to my head. Indeed, if the much more massive plane fell faster than both plastic Albert and myself, we'd both bump into the ceiling! The fact that we all floated around together is a beautiful demonstration of the fact that all objects, no matter what their mass, fall at the same rate in a gravitational field.

This simple fact inspired Albert Einstein to construct his geometric theory of gravitation, called General Relativity, which to this day is the most accurate theoretical description of gravity that we possess. We shall get to Einstein's beautiful theory later on, and in doing so we'll arrive at a very simple explanation of why everything falls at the same rate, and why gravity can be removed by the act of falling●

Gravity holds the water in our oceans and hugs the atmosphere close to the planet. It's the reason why the rain falls and the rivers flow; it powers the ocean currents and drives the world's weather; it's why volcanoes erupt and earthquakes tear the land apart. Yet gravity also plays a role on an even grander stage. Across the Universe, from the smallest speck of dust to the most massive star, gravity is the great sculptor that created order out of chaos.

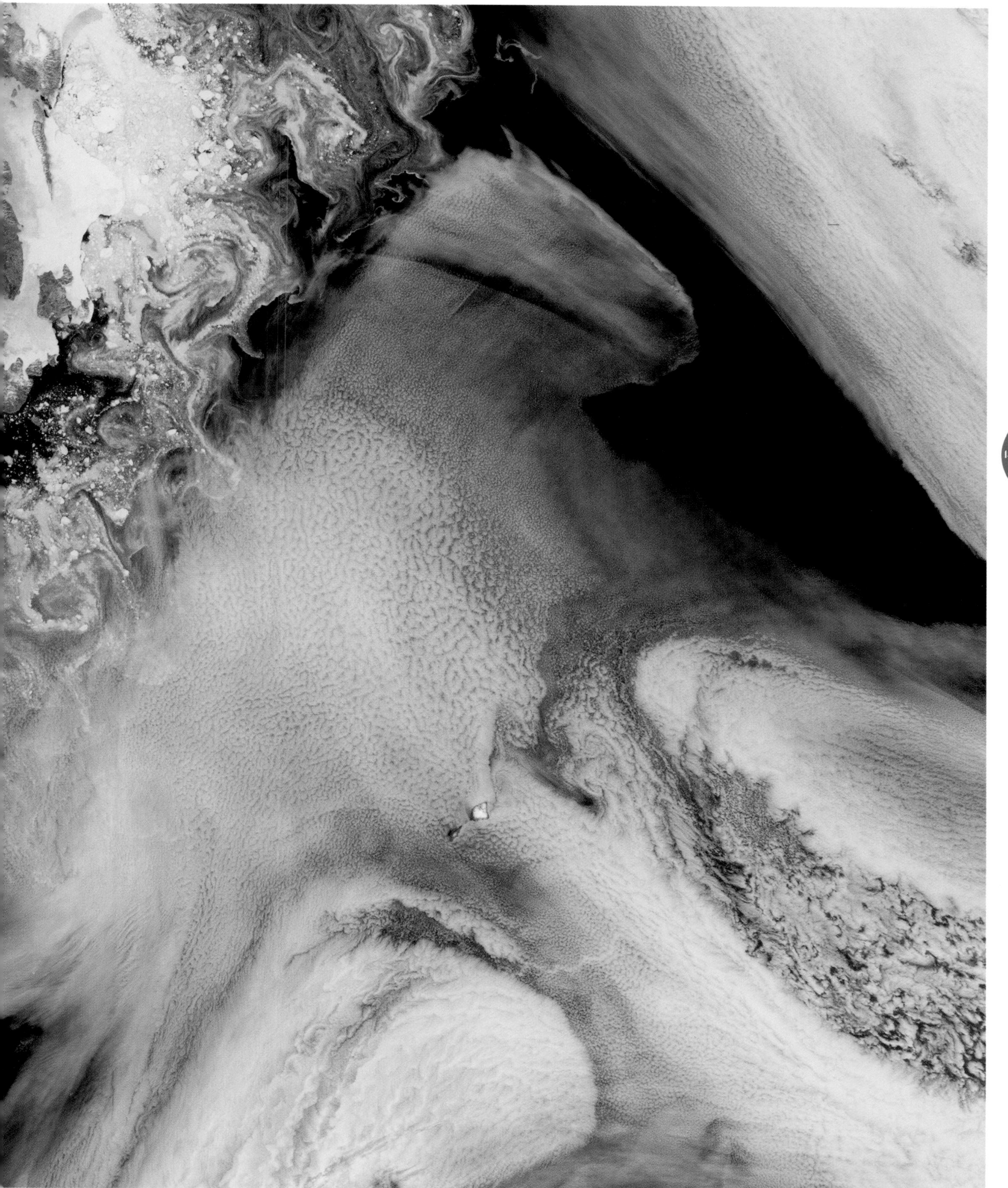

THE INVISIBLE STRING

Everything in the Cosmos is subject to the force of gravity. From the manmade satellites that rotate around our planet creating the technological infrastructure of the twenty-first century, to the orbit of our only natural satellite – the Moon – which journeys around Earth every 27.3 days, it is gravity that provides the invisible string to guide them on their path. The journey of every planet, moon, ball of rock and mote of dust in our solar system is guided by gravity; from the 365-day trip our planet takes around the Sun to each of the orbits of the seven planets and 166 known moons in our neighbourhood. Beyond our solar system, gravity continues to conduct the flow of the Universe, with everything affected by the gravitational pull of something else, no matter how tiny or how massive.

Our solar system orbits around the centre of the Milky Way Galaxy, a place dominated by a supermassive black hole, the heart of a swirling system of over 200 billion gravitationally bound stars. And even this vast, rotating structure isn't where the merry-go-round of the Universe ends, because even the galaxies are steered through the vast Universe by the action of gravity.

LEFT: The Virgo Supercluster of galaxies is a good example of how gravitational pull exerts itself. This cluster of galaxies has a gravitational pull on the Local Group of galaxies that surround our Milky Way Galaxy.

TOP: The supermassive black hole at the centre of the Milky Way Galaxy, Sagittarius A*, is the heart of a swirling system of over 200 billion stars which are gravitationally bound.

ABOVE: The elliptical galaxy M87 is located at the centre of the Virgo Cluster. This huge galaxy includes several trillion stars, a supermassive black hole, and a family of 15,000 globular star clusters which may have been graviationally pulled from nearby dwarf galaxies.

Beyond our solar system, gravity continues to conduct the flow of the Universe, with everything affected by the gravitational pull of something else, no matter how tiny or how massive.

Our galaxy is part of a collection of galaxies called the Local Group – a cluster of over 30 galaxies named by the American astronomer Edwin Hubble in 1936. Over ten million light years across, this vast dumbbell-shaped structure contains billions and billions of stars, including the trillion stars that make up our giant galactic neighbour, Andromeda. Just as the Moon orbits Earth, Earth orbits the Sun, and the Sun orbits the Milky Way, so the Local Group orbits its common centre of gravity, located somewhere in the 2.5 million light years between the two most massive galaxies in the group: our Milky Way and Andromeda. But even this giant community of galaxies isn't the largest known gravitationally bound structure. As you sit reading this book, gravity is taking you on an extraordinary ride. Not only are you spinning around as Earth rotates once a day on its axis, not only are you orbiting at just over 100,000 kilometres (62,137 miles) per hour around the Sun, not only are you rotating around the centre of our galaxy at 220 kilometres (136 miles) per second, and not only is the entire Milky Way tearing around the centre of gravity of the Local Group at 600 kilometres (372 miles) per second, but we are also part of even an grander gravitationally driven cycle.

The Local Group is part of a much larger, gravitationally bound family called the Virgo Supercluster – a collection of at least 100 galaxy clusters. Nobody is sure how long it takes our Local Group to journey around the Virgo Supercluster; vast beyond words, stretching over 110 million light years, it is, even so, only one of millions of superclusters in the observable Universe. It is now thought that even superclusters are part of far larger structures bound together by gravity, known as galaxy filaments or great walls. We are part of the Pisces-Cetus Supercluster Complex.

Gravity's scope is unlimited, its influence all-pervasive at all distance scales throughout the entire history of the Universe. Yet, perhaps surprisingly, given its colossal reach and universal importance, it is the first force that we humans understood in any detail ◉

THE APPLE THAT NEVER FELL

The history of science is littered with examples of circumstance and serendipity leading to the greatest discoveries, which is why curiosity-driven science is the foundation of our civilisation. Among the most celebrated is the convoluted story of Newton's journey to his theory of gravity – the first great universal law of physics.

The Great Plague of 1665 was the last major outbreak of bubonic plague in England, but also the most deadly. Over one hundred thousand people are thought to have died the hideous death that accompanied the rodent-borne illness. London was the epicentre of the outbreak, but even then the matrix of connections between the capital and the rest of the country caused the disease to spread rapidly across England. Extreme and often useless measures were taken to prevent its spread, from the lighting of fires to cleanse the air to the culling of innocent dogs and cats. Infected villages were quarantined and schools and colleges closed. One place affected was Trinity College Cambridge, and one of the students to take a leave of absence in the summer of 1665 was Isaac Newton.

Newton was twenty-two years old and newly graduated when he left plague-ridden Cambridge to return to his family home in Woolsthorpe, Lincolnshire. He took with him a series of books on mathematics and the geometry of Euclid and Descartes, in which he had become interested, he later wrote, through an astronomy book he purchased at a fair. Although by all accounts he was an unremarkable student, his enforced absence allowed him time to think, and his interest in the physical world and the laws underpinning it began to coalesce. Over the next two years his private studies laid the foundations for much of his later work in subjects as diverse as calculus, optics and, of course, gravity. On returning to Cambridge in 1667 he was elected as a fellow, and became the Lucasian Professor of Mathematics in October 1670 (a post recently held by Stephen Hawking and currently held by string theorist Michael Green – both of whom continue to work on the problem of the nature of gravity). Newton spent the next twenty years lecturing and working in a diverse range of scientific and pseudo-scientific endeavours, including alchemy and predictions of the date of the apocalypse. The economist John Maynard Keynes said of Newton that he was not 'the first in the age of reason, but the last of the magicians'. This is not entirely accurate, but then what can one reasonably expect from an economist? Newton lived on the cusp of pre-scientific times and the modern age and did more than most to usher in the transition. His greatest contribution to modern science was the publication in 1687 of the *Philosophiæ Naturalis Principia Mathematica*, otherwise known as the *Principia*. This book contains an equation that describes the action of gravity so precisely that it was used to guide the Apollo astronauts on their journey to the Moon. It is beautiful in its simplicity

THE EFFECT OF GRAVITY ON THE MOVEMENT OF PLANETS

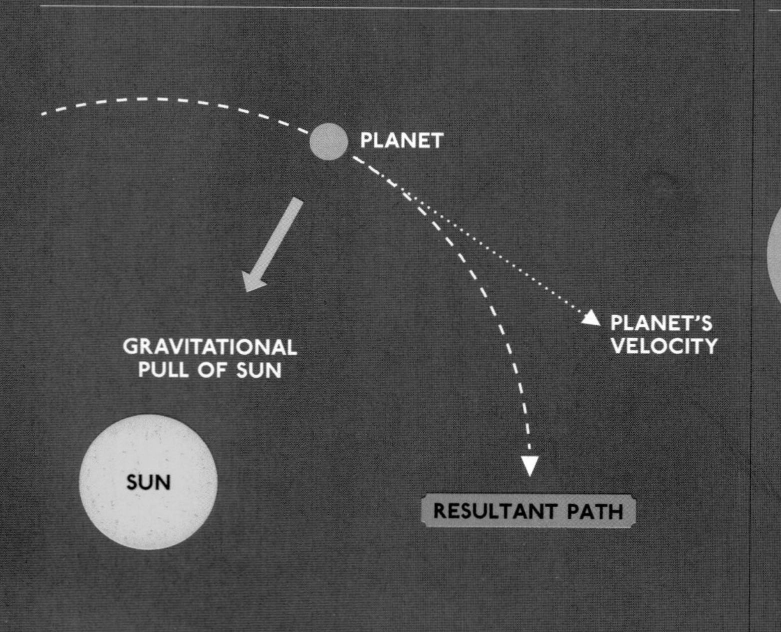

NEWTON'S LAW OF UNIVERSAL GRAVITATION

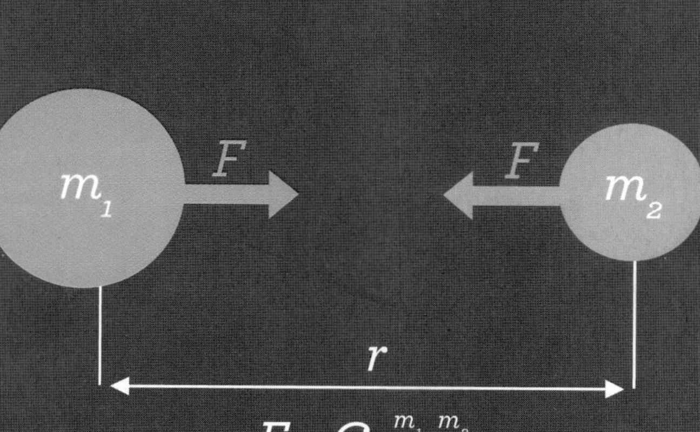

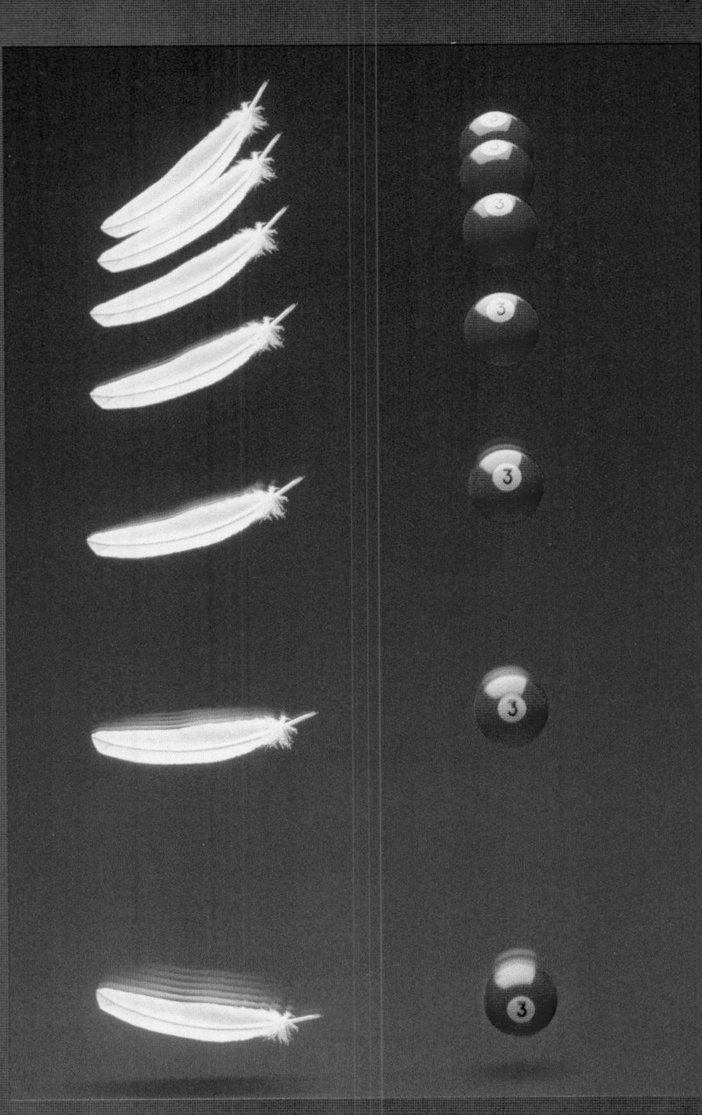

LEFT: This time-lapse image neatly illustrates the concept of gravity. The feather and ball are here seen falling at the same speed in a vacuum, proving that any two objects of different mass will accelerate at identical rates when at the same gravitational potential. The reason that this does not happen on Earth is because of the air resistance that is present, which is, of course, absent in a vacuum. This principle was also proved correct when an Apollo astronaut dropped a feather and a hammer on the Moon (which has no atmosphere) and saw them fall at the same rate.

and profound in its application and consequences for scientific thought.

$$F = G\frac{m_1 m_2}{r^2}$$

This is the mathematical expression of Newton's Law of Universal Gravitation. In words, it says that the force (F) between two objects is equal to the product of their masses (m_1 and m_2), divided by the square of the distance between them. G is a constant of proportionality known as the gravitational constant; its value encodes the strength of the gravitational force: The force between two one-kilogramme masses, 1 metre (3 feet) apart, is 6.67428 x 10^{-11} newtons – that's 0.0000000000667428 N, which is not a lot. For comparison, the force exerted on your hand by a 1kg bag of sugar is approximately 10 N. In other words, the gravitational constant G is 6.67428 x 10^{-11} N $(m/Kg)^2$. The reason why G is so tiny is unknown and one of the greatest questions in physics; the electromagnetic force is 10^{36} times stronger – that's a factor of a million million million million million million.

There are many reasons why Newton's Law of Universal Gravitation is beautiful. It is universal, which means it applies everywhere in the Universe and to everything not in the vicinity of black holes, too close to massive stars or moving close to the speed of light. In these cases, Einstein's more accurate theory of General Relativity is required. For planetary orbits around stars, orbits of stars around galaxies and the movements of the galaxies themselves, it is more than accurate enough. It has also applied at all times in the Universe's history beyond the first instants after the Big Bang. This is not to be taken for granted, because the law was derived based on the work of Johannes Kepler and the observations of Tycho Brahe, who were concerned only with the motion of the planets around the Sun. The fact that a law that governs the clockwork of our solar system is the same law that governs the motion of the galaxies is interesting and important. It is the statement that the same laws of physics govern our whole universe, and Newton's law of gravitation was the first example of such a universal law.

It is also profoundly simple. That the complex motion of everything in the cosmos can be summed up in a single mathematical formula is elegant and beautiful, and lies at the heart of modern fundamental science. You don't need to sit down with a telescope every night and use trial and error to find the positions of the planets and moons of the Solar System. You can work out where they will be at any point in the future using Newton's simple equation, and this applies not just to our solar system, but also to every solar system in the Universe. Such is the power of mathematics and physics.

Newton found that gravity is a force of attraction that exists between all objects, from the tiny immeasurable force of attraction between two rocks on the ground to the rather larger force that each and every one of us is currently experiencing between our bodies and the massive rock upon which we are stood. With a mass of almost 6 milllion million million million kilogrammes, the force between all of us and our planet is strong enough to keep our feet on the ground. On the scale of planets, however, gravity can do much more than simply keep them in orbit and hold things on the ground; it can sculpt and shape their surfaces in profound and unexpected ways ●

THE GRAND SCULPTURE

Fish River Canyon in the south of Namibia is one of the world's great geological features, second only in scale to the Grand Canyon in Arizona, at over 160 kilometres (99 miles) long, 26 kilometres (16 miles) wide and half a kilometre (a third of a mile) deep in places. Like the Grand Canyon, the movement of tectonic plates or volcanic action did not create this scar in Earth's crust; instead it stands testament to the erosive power of water. The Fish River is the longest river in Namibia, running for over 650 kilometres (403 miles). Despite only flowing in the summer, over millennia it has slowly but forcefully gouged the canyon out of solid rock. This takes energy, and that energy ultimately comes from the Sun as it lifts water from the oceans and deposits it upstream in the highlands to the north. Once the rain begins to fall, gravity takes over. The highlands around the source of the Fish River are at an elevation of over a thousand metres above sea level. When the rain lands on the ground at this elevation, every water droplet stores energy in the form of gravitational potential energy. There is a simple equation that says how much energy each drop has stored up:

$$U = mgh$$

U is the amount of energy that will be released if the drop falls from height (h) above sea level down to sea level, m is the mass of the drop and g is the now-familiar acceleration due to gravity – 9.81 m/s^2.

Every droplet of water raised high by the heat of the Sun has energy, due to its position in Earth's gravitational field, and this energy can be released by allowing the water to flow downwards to the sea. Some of this energy is available to cut deep into Earth's surface to form the Fish River Canyon.

The strength of Earth's gravitational field therefore has a powerful influence on its surface features. This is not only visible in the action of falling, tumbling water, but in the size of its mountains. On Earth, the tallest mountain above sea level is Mount Everest; at almost 9 kilometres (5.5 miles), it towers above the rest of the planet. But Everest is dwarfed by the tallest mountain in the Solar System which, perhaps at first sight surprisingly, sits on the surface of a much smaller planet. Around 78 million kilometres (48 million miles) from Earth, Mars is similar to our planet in many ways. Its surface is scarred by the action of water that once tumbled from the highlands to the seas, dissipating its gravitational potential energy as it fell, although today, the water has left Mars. The planet is only around 10 per cent as massive as Earth, though, so its gravitational pull is significantly weaker, and this is one of the reasons why Mars was unable to hang on to its atmosphere, despite being further away from the Sun. The possibility of liquid water flowing on the Martian surface vanished with its atmosphere, leaving the red planet to an arid and geologically dead future, but Mars's lower surface gravity has a surprising consequence for its mountains.

Towering over every other mountain in the Solar System is the extinct volcano, Olympus Mons. Rising to an altitude of around 24 kilometres (15 miles), it is almost the height of three Mount Everests stacked on top of each other. The fact that a smaller planet has higher mountains is not a coincidence; it is partly down to environmental factors such as the rate of erosion and the details of the planet's geological past, but there is also a fundamental limit to the height of mountains on any given planet: the strength of its surface gravity. Mars has a radius approximately half that of Earth's, and since it is only 10 per cent as massive, a little calculation using Newton's equation will tell you that the strength of the gravitational pull at its surface is approximately 40 per cent of that on our planet. This changes everything's weight.

Here on Earth we don't often think about the difference between mass and weight, but the distinction is very real. The mass of something is an intrinsic property of that thing – it is a measure of how much stuff the thing is made of. This doesn't change, no matter where in the Universe the thing is placed. In Einstein's Theory of Special Relativity, the rest mass of an object is an invariant quantity, which means that everyone in the Universe, no matter where they are or how they are moving, would measure the same value for the rest mass.

Weight is different. For one thing, it is not measured in kilogrammes, it is measured in the units of force – newtons. This is easy to understand if you think about how you would measure your weight. When you stand on bathroom scales, they measure the force being exerted on them by you; you can see this by pressing down on them – the harder you push, the greater the weight reading. The force you are exerting on the scales is in turn dependent on the strength of Earth's gravity. This should be obvious; if I had taken the scales up

BELOW AND PREVIOUS SPREAD: The Fish River Canyon in southern Namibia is one of the world's greatest geological sites, and a spectacular example of how the effects of climate and gravity can impact on the structure of Earth's surface.

RIGHT: The immense Olympus Mons can exist on Mars because the planet has 40 per cent of Earth's gravitational pull. However, move this extinct volcano to our planet and it would sink into the ground because of its enormous weight.

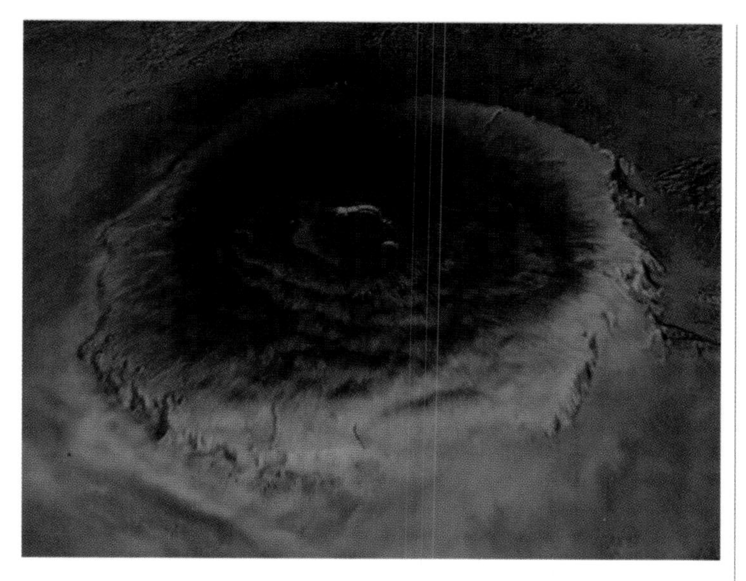

If you took Olympus Mons and stuck it on Earth ... it would weigh around two and a half times as much as it does on Mars ... A planet the size of ours cannot sustain a mountain of this size – it would weigh too much.

in the Vomit Comet and tried to stand on them, they wouldn't have read anything because I would have been floating above them – hence the word 'weightless'. In symbols, the weight of something on Earth is defined as:

$$W=mg$$

W is weight, m is the thing's mass, and g is the familiar measure of Earth's gravitational field strength – 9.81 m/s^2 – with a couple of caveats that we'll get to below! (For absolute accuracy, the correct definition of weight is the force that is applied on you by the scales to give you an acceleration equal to the local acceleration due to gravity – i.e. the force the scales exert on you to stop you falling through them.) So, here on Earth a human being with a mass of 80kg weighs 785 newtons; on Mars, the same 80-kg person would weigh approximately 295 newtons.

So your weight depends on a few things; one is your mass, another is the mass of the planet you are on. Your weight would also change if you were accelerating when you measured it, which is another manifestation of the equivalence principle. So, if you took Olympus Mons and stuck it on Earth, then as well as dwarfing every other mountain on the planet, it would also weigh around two and a half times as much as it does on Mars. This enormous force would put its base rock under such intense pressure that it would be unable to support the mountain, so it would sink into the ground. A planet the size of ours cannot sustain a mountain the size of Olympus Mons – it would weigh too much. The highest mountain on Earth, as measured from its base, is Mauna Kea, the vast dormant volcano on Hawaii. It is over one kilometre (half a mile) higher than Everest, and it is gradually sinking. So Mauna Kea is as high as a mountain can be on our planet, and this absolute limit is set by the strength of our gravity.

The definition of weight can get a bit convoluted, and we mentioned that there are caveats to the rule of thumb that your weight on Earth is 9.81 times your mass. One problem is that the strength of Earth's gravity varies slightly at every point on its surface. The most obvious effect is altitude; on the edge of the Fish River Canyon I would weigh slightly less than I would if I stood on the canyon floor. That's because at the top of the canyon I am further from the centre of Earth than I would be at the bottom, so the gravitational pull I feel is weaker. Earth is also not uniformly dense – some areas of Earth's surface and subsurface are made of more massive stuff than others, which also affects the local gravitational field. To complicate matters further, Earth is spinning, which means that you are accelerating when you stand on its surface, which means that the strength of gravity you feel changes in accord with the equivalence principle; this acceleration increases as you go towards the Equator, reducing the gravitational acceleration you feel there. Earth bulges out at the Equator because it is spinning, which weakens the gravitational pull there still further. The upshot of all this is that you weigh approximately 0.5 per cent less at the North and South Poles than you do at the Equator. The effects of the varying density of Earth's subsurface and the presence of surface features on Earth's gravitational field have been measured to extremely high precision and presented as a map known as the geoid ◉

Towering over every other mountain in the Solar System is the extinct volcano, Olympus Mons. It is almost the height of three Mount Everests stacked on top of each other. The fact that a smaller planet has higher mountains is not coincidence; it is partly down to environmental and geological factors, but there is also a fundamental limit to the height of mountains on any given planet; the strength of its surface gravity. Mars has a gravitational pull at its surface of approximately 40 per cent of that on our planet.

THE GEOID

BELOW: Data collected by the GOCE satellite between November and December 2009 is here used to create a map of the tiny variations in Earth's gravity field across the globe. These maps provide invaluable information for oceanographers, hydrologists and geologists in order to create accurate climate models for our planet.

RIGHT: The geoid helps us to understand unseen structures on our planet, such as here in Iceland where magma is welling upwards from Earth's mantle, affecting the gravitational field there. In this image, taken in May 2010 from a NASA satellite, the Icelandic volcano Eyjafjallajökull can be seen erupting.

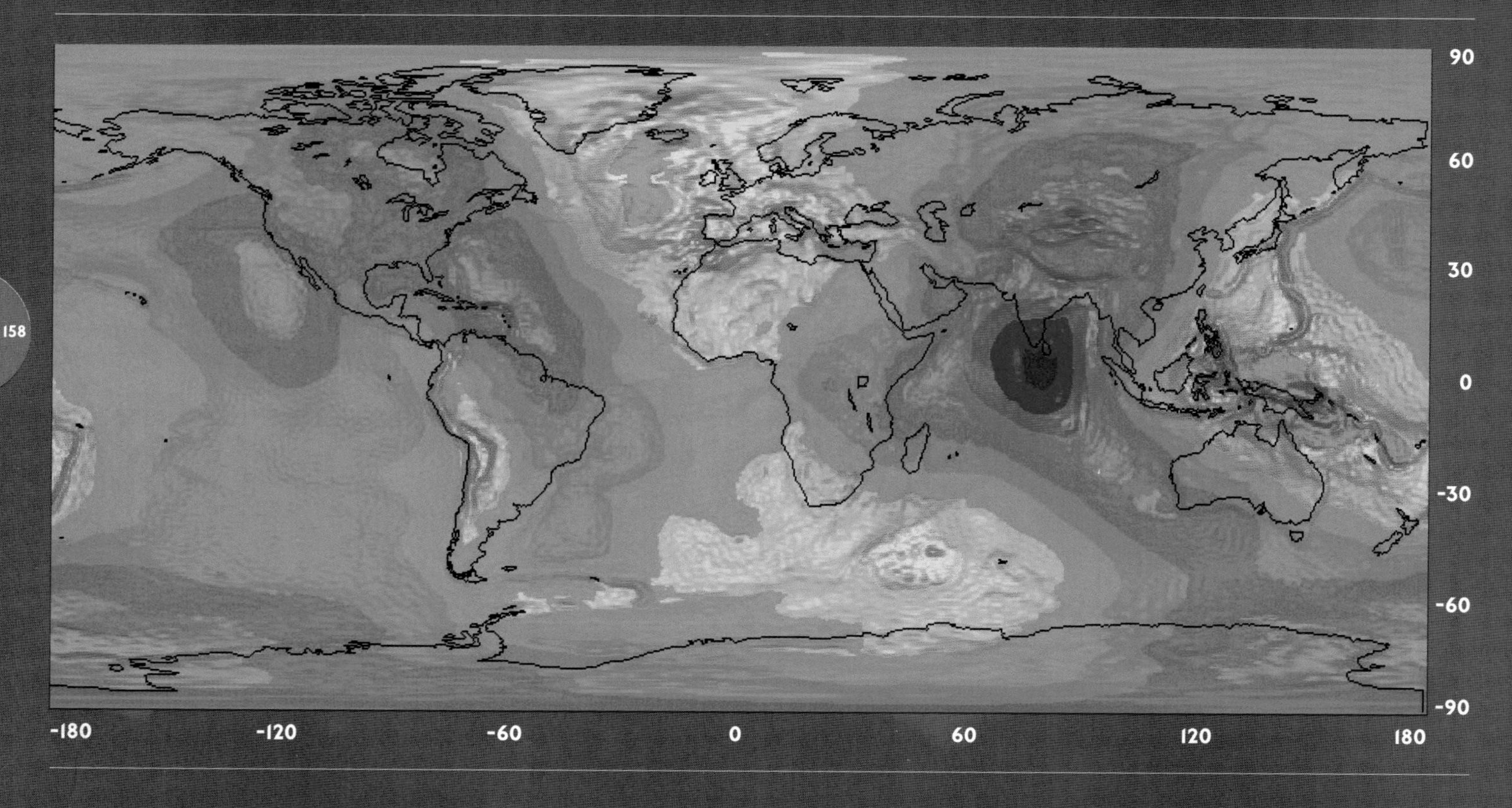

This picture of Earth's gravitational field was taken by a European Space Agency satellite, GOCE, which was launched in March 2009. GOCE is equipped with three ultra-sensitive accelerometers, arranged so that they respond to very tiny changes in the strength of Earth's gravitational field as the satellite orbits. Skimming the edge of Earth's atmosphere at an altitude of 250 kilometres (155 miles), GOCE spent two months gathering the data to create this extraordinary image. It's the first time the strength of gravity across the globe has been mapped this accurately. The blue patches indicate areas that have a weak gravitational field, the green are average and the red are places where it is stronger. The reason for these fluctuations is the density of the rocks below Earth's surface and the presence of features such as mountains or ocean trenches. More technically, the picture is presented as an equipotential surface, which means that if Earth were entirely covered in a single ocean of water, this picture would correspond to the water height at every point.

Looking at this map, it is clear that Iceland has a higher gravitational field strength than that of England. These changes are imperceptible to us, but it means that I would weigh slightly less standing at the same altitude in Manchester than I would in Reykjavik. This map was not made to show the trivial distinctions in a traveller's weight, of course; the unparalleled level of detail will enable a deeper understanding of how our planet works, because this data is a high-precision geological tool. One particular benefit will be for oceanographers; because the map defines the baseline water surface in the absence of tides, winds and currents, it is critical to understanding the factors that determine the movement of water across the oceans of our planet. This is a very important part of understanding and predicting the way energy is transferred around our planet, which is in turn an important factor in generating accurate climate models.

The geoid therefore reveals a vast amount of detailed information about the structure of our planet, just from measuring the strength of its gravity. As far as the actual height of the ocean surface is concerned, however, the most influential factor of all is not shown: the Moon ◉

THE TUG OF THE MOON

Many of the planets that exist in our solar system have families of moons; from the sixty-three satellites of Jupiter, to the thirteen moons of Neptune, and to the two tiny misshapen moons of Mars. Our planet has only a single moon; it is our constant companion, with which we have travelled through space for almost four and a half billion years.

RIGHT: The elusive far side of the Moon, which was eventually first photographed in 1959 by the Soviet Luna 3 probe.

No other planet in our solar system has a moon as large as ours in relation to its parent planet. Orbiting only 380,000 kilometres (236, 000 miles) from Earth, it is a quarter of the Earth's diameter, making it the fifth-largest moon in the Solar System after Titan, Ganymede, Callisto and Io – although of course their parent planets, Jupiter and Saturn, are significantly larger than Earth. This makes the Earth and Moon close to being a double-planet system. The current best theory for the formation of our moon is that it was created around 4.5 billion years ago when a Mars-sized planet, which has been named Theia, crashed into the newly formed Earth, blasting rock into orbit which slowly condensed into the lunar structure that we see today. The evidence for this theory is partly that the Moon has a very similar composition to that of Earth's outer crust, although it is much less dense because it has a significantly smaller iron core. This is what would be expected if the Theia/Earth collision was a glancing blow, leaving the Earth's iron core intact and so reducing the relative amount of iron in the Moon. This in turn means that the Moon's gravitational field is much weaker than ours. When Neil Armstrong took his small step onto the Moon, he weighed just 26 kilogrammes (58 pounds), despite the fact that he was wearing a space suit that had weighed 81 kilogrammes (180 pounds) on its own on Earth – this is all because the Moon's gravitational field strength is approximately one-sixth of Earth's. Despite this relatively weak gravitational pull, however, the Moon still has a profound effect on our planet.

Because of the Moon's proximity to our planet, its gravitational pull varies significantly from one side of Earth to the other. The illustration (right) shows the net gravitational force exerted at each point on Earth by the Moon, as seen by someone sitting at Earth's centre, after Earth's own gravitational field has been subtracted away. What remains is a net gravitational force pulling the side of the Earth that is facing the Moon towards the Moon, as you might expect. But there is also a net force pulling the opposite side of Earth away from the Moon. Notice also that at right angles to the position of the Moon, the lunar gravity actually adds to the Earth's gravitational pull and squashes everything! This is the origin of the tides; because water is easier to stretch than the rock that forms the ocean floor, the water in the oceans bulges outwards relative to the ground beneath the Moon and on the opposite side of Earth to the Moon. The difference in water heights is only a few metres, but can be much higher depending on the shape of the shoreline. It's worth mentioning that there are also tides in the rocks of Earth's surface; gravity doesn't just affect water! But rocks are very rigid, and so don't stretch much. The surface of Earth does, however, rise and fall by a few centimetres due to tidal effects. As Earth rotates beneath the tidal bulge raised in the oceans, the distorted water surface sweeps past the shorelines and we experience two high and low tides per day.

Next time someone starts trying to tell you that we are made of water and therefore the Moon must have an influence on us, you will now be justified in having a strange, blank and perhaps slightly pitying expression on your face for two reasons. One is that because the tides are a differential effect (that is to say they depend on the change in the strength of the Moon's gravity across the diameter of Earth), the tidal effect on you is utterly insignificant and makes no difference to you at all because the difference in the Moon's gravitational force across something the size of your body is negligible. Secondly, it has got nothing at all to do with water in any case!

Gravity is always a two-way street – just as the Moon raises tides on Earth, so Earth must cause tides to sweep across the surface of the Moon.

The relationship between the Earth and the Moon is not just a one-way street; just as the Moon's gravity has transformed our planet, so in turn Earth has transformed its neighbour.

Throughout human history, half of the Moon's surface remained hidden from view, and it wasn't until 1959, when the Soviet Luna 3 probe photographed the far side of the Moon for the first time, that we caught our first glimpse of this hidden landscape. Nine years later, the astronauts on board Apollo 8 became the first humans to leave Earth's orbit and the first human beings to directly observe the far side of the Moon with their own eyes. The reason only one side of the Moon faces Earth, appearing frozen in time and unchanging in the seemingly ever-moving night sky, is down to the tidal effects.

Billions of years ago, the view of our satellite from Earth would have been very different. In its childhood, the Moon rotated much faster, and both sides of its surface would have been visible from Earth. From the moment of its birth, the Moon felt the tug of Earth's gravity – a force that would have been even greater than it is today because the Moon was also closer to Earth.

LUNAR GRAVITY DIFFERENTIAL FIELD
The lunar gravity differential field at Earth's surface is known as the tide-generating force. This is the primary mechanism that drives tidal action and explains two equipotential tidal bulges, accounting for two daily high waters.

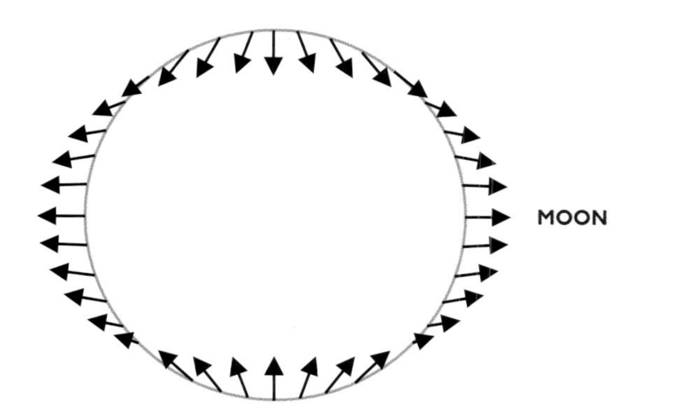

THE EFFECT OF TIDAL LOCKING ON THE EARTH AND MOON
As the Earth–Moon system moves towards being perfectly tidally locked, the Moon is gradually drifting away from Earth.

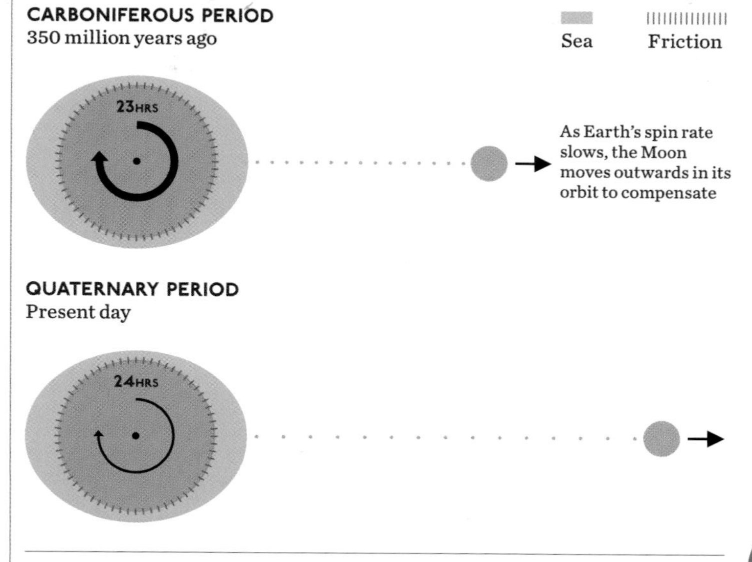

LEFT AND ABOVE: The Moon has a visible effect on our oceans. The combination of the gravitational pulls of the Moon and of Earth squashes everything, which in turn creates tides.

A glance at Newton's Law of Universal Gravitation will tell you that gravity is always a two-way street – just as the Moon raises tides on Earth, so Earth must cause tides to sweep across the surface of the Moon. These tides are not in water, of course, but in the solid rock of the lunar surface. In an amazing piece of planetary heavy lifting, the Moon's crust would have been distorted by up to 7 metres (22 feet)!

This giant tidal bulge sweeping across the Moon had an interesting effect. As the Moon turned beneath the giant parent planet hanging in the lunar sky, the rock tide was dragged across its surface, but the rising of the tide isn't instantaneous; it takes time for the surface of the Moon to respond to the pull of the Earth. During that time, the Moon will have rotated a bit, carrying the peak of the rock tide with it. The tidal bulge will therefore not be in perfect alignment with Earth, but slightly ahead of it. Earth's gravity acts on the misshapen Moon in such a way that it tries to pull it back into sync; in other words, it works like a giant brake. Over time, this effect, known as tidal locking, gradually synchronizes the rotation rate of the Moon with its orbital period, effectively meaning that the tidal bulge can remain in exactly the same place on the Moon's surface beneath Earth and doesn't have to be swept around.

The Moon is now almost, but not quite, tidally locked to Earth, which means that it takes one month to rotate around on its axis and one month to orbit Earth. So there's no dark side of the Moon – the side we can't see gets plenty of sunlight, it's just a side that perpetually faces away from Earth. The Earth–Moon system is in fact still evolving towards being perfectly tidally locked, and one interesting consequence of this is that the Moon is gradually drifting further and further away from Earth at a rate of just under 4 centimtres (1.5 inches) per year.

The power of gravity is not just in its ability to reach across the empty wastes of space and shape the surface of planets and moons; gravity also has the power to create whole new worlds, and we can see the process of that creation frozen in time in the sky, every day and every night ◉

THE FALSE DAWN

It is one of the strangest lights that appears in our night sky; a light that for centuries has puzzled those who have witnessed its glow, fooling them into thinking that a new day was arriving. The Prophet Muhammed called it the false dawn and warned the followers of Islam not to confuse it with the real dawn when setting the timing of daily prayers.

This magical glow that appears on the horizon just before sunrise and just after sunset has nothing to do with the arrival or departure of our star; instead it is a ghostly reminder of our world's origins and the power of gravity. It is the Zodiacal light; a wispy, whitish glow that appears to form a rough triangular shape rising from the horizon. The Italian astronomer Giovanni Cassini first investigated this strange phenomenon in 1683. The ethereal light perplexed many scientists of the age, and a common explanation was that the light came from the atmosphere of the Sun as it rose above the horizon before the Sun itself. It was Nicolas Fatio de Duillier, one of Cassini's students, who finally explained its origin, and in doing so he provided a first glimpse of the origin of the planets and moons in our solar system.

The story of the Zodiacal light can be traced back five billion years to the origins of our solar system. Back then, there was no Sun, nor any planets or moons; there was only a cloud of gas and dust, the building blocks of everything we now call home. Everything that makes up our solar system was contained in an enormous irregular cloud floating through space. It is thought the explosion of a nearby star sent a shockwave through the cloud, creating small fluctuations in density. It also imparted rotation. The denser regions had slightly more gravitational pull than the less dense regions, so they began to grow, and the largest one became the Sun. In its earliest days the Solar System would have been planet-less; surrounding the young Sun was a spinning disc of matter, a protoplanetary disc. Over time, the minute particles of dust in the disc collided and clumped together, and large objects the size of small asteroids, known as planetesimals, would have formed by chance. Once the larger planetesimals were big enough to have significant gravity, they began to sweep up the matter close to them and their growth accelerated. Roughly one hundred million years later, the largest planetesimals evolved into the planets and moons we see today.

However, not all this matter from the primordial cloud became a planet or moon. Out in the solar system beyond Mars there should be another planet, but a gravitational tug of war between Jupiter and the Sun stops it forming. Now, instead of a ninth planet, there is a band of dust and debris – the asteroid belt. Normally there is no way of seeing the asteroid belt from Earth with the naked eye – it's just too far away and the asteroids are too small – but collisions within the asteroid belt produce dust, and that is the secret behind the false dawn. The faint glow of the Zodiacal light after sunset and before sunrise is caused by sunlight reflecting off the debris of a failed planet; a remnant of the early Solar System and a beautiful, glimmering reminder of our origins ◉

BELOW: The wispy, whitish glow that appears on the horizon before sunrise and just after sunset was a subject of great debate among scientists for centuries. This Zodiacal light, as it is known, is in fact the debris that remains after collisions within the asteroid belt caused by a gravitational tug of war in the Solar System.

RIGHT: Theoretically, another planet should have formed from the primordial dust in the Solar System beyond Mars; however, the conflicting gravitational forces between the Sun and Jupiter prevent this happening, resulting in a band of dust and debris known as the asteroid belt.

Normally there is no way of seeing the asteroid belt from Earth with the naked eye – it's just too far away and the asteroids are too small – but collisions within the asteroid belt produce dust, and that is the secret behind the false dawn.

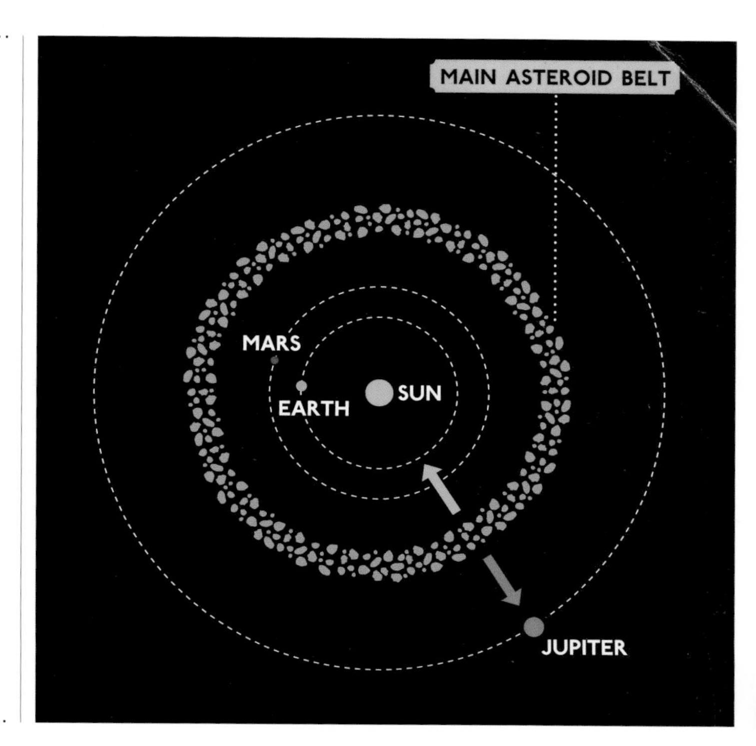

RIGHT: 'The Blue Marble' is perhaps one of the most famous photographs ever taken of Earth, and has inspired numerous images since. The photograph, taken by the Apollo 17 crew on their 1972 journey to the Moon, made history as the first true-colour image of our planet which showed Earth in unprecedented detail.

THE BLUE MARBLE

Even the most dogmatic flat-Earther would have a problem explaining away 'The Blue Marble'. This photo, taken by the astronauts on board Apollo 17 during its journey to the Moon on 7 December 1972, has caused some to speculate that this beautiful picture of our fragile world is perhaps the most distributed image in human history. But why is Earth a sphere? Actually, why are all planets and all stars spherical?

As we've discussed, we know that planets and stars are formed by the gravitational collapse of clouds of dust. You could say that the force of gravity pulls everything together,

'The Blue Marble' ... photo has caused some to speculate that this beautiful picture of our fragile world is perhaps the most distributed image in human history.

which is one way of looking at it, but another way of saying the same thing is that all the little particles in the primordial cloud of dust had gravitational potential energy, because they were all floating around in each other's tiny gravitational fields. Just like the water droplets that fell as rain high up in the mountains above the Fish River Canyon, these particles would all try to fall 'downhill' to minimise their gravitational potential energy. This leads us to a very general and very deep principle in physics, and you can pretty much explain everything that happens in the Universe by applying it: things will minimise their potential energy if they can find a way of doing so. So, you could answer the question 'why does a ball roll down a hill?' by saying that the ball would have lower gravitational potential energy at the bottom of the hill than the top, so it rolls down. You could also, of course, say that there is a force pulling the ball down the hill. Physicists often work with energies rather than forces, and the two languages are interchangeable.

With a collapsing cloud of dust, the shape that ultimately forms will therefore be the shape that minimises the gravitational potential energy. The shape must be the one that allows everything within the cloud to get as close to the centre of it as it possibly can, because anything that is located further away from the centre will have more gravitational potential energy! So, the shape that ensures that everything is as close to the centre as possible is, naturally, a sphere, which is why stars and planets are spherical ◉

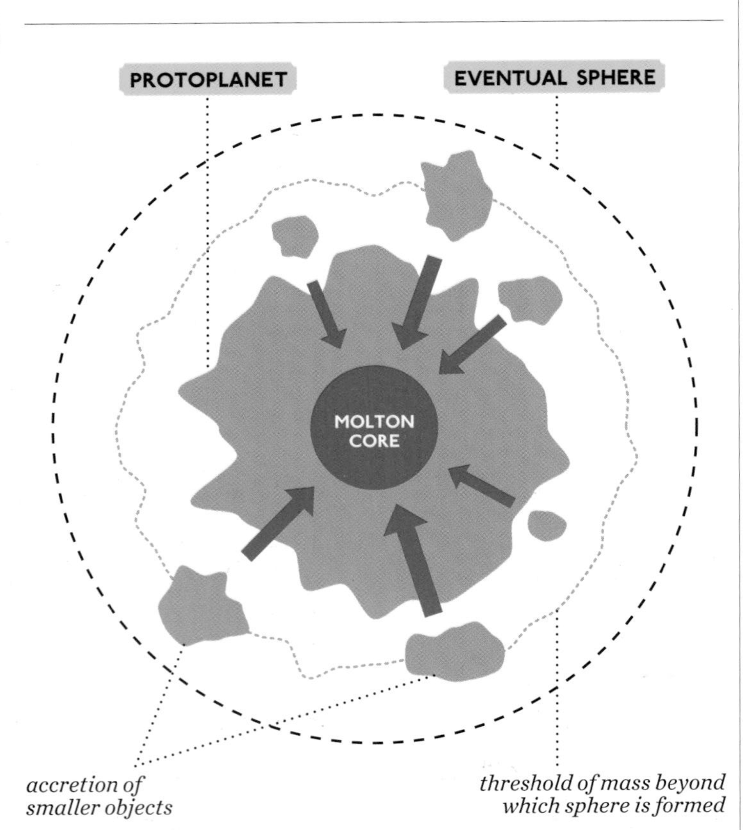

VERY LARGE ARRAY

BELOW: A very large array indeed – the 27 dishes on the Plains of San Augustin are an impressive sight, stretching into the horizon. Through these, the radio astronomy observatory can take some even more impressive images.

In the US state of New Mexico, on the Plains of San Augustin between the towns of Magdalena and Datil, lies one of the most spectacular and iconic observatories on the planet. The Very Large Array (VLA) is a radio astronomy observatory consisting of 27 identical dishes, each 25 metres (82 feet) in diameter, arranged in a gigantic Y shape across the landscape. Although each dish works independently, they can be combined together to create a single antenna with an effective diameter of over 36 kilometres (22 miles). This allows this vast virtual telescope to achieve very high-resolution images of the sky at radio wavelengths.

Radio astronomy has a history dating back to the 1930s, when the astronomer Karl Jansky discovered that the Universe could be explored not just through the visible part of the electromagnetic spectrum, but also through the detection of radio waves. Over a period of several months, Jansky used an antenna that looked more like a Meccano set than the VLA to record the radio waves from the sky. He initially identified two types of signal: radio waves generated by nearby thunderstorms, and radio waves generated by distant thunderstorms. He also found a third type, a form of what he thought was static. The interesting thing about the static was that it seemed to rise and fall once a day, which suggested to Jansky that it consisted of radio waves being generated from the Sun, but then over a period of weeks the rise and fall of the static deviated from a 24-hour cycle. Jansky could rotate his antennae on a set of Ford Model T tyres to follow the mysterious signal, and he

soon realised the brightest point was not coming from the direction of the Sun, but from the centre of the Milky Way Galaxy in the direction of the constellation of Sagittarius.

Coinciding with the economic impact of the Great Depression, Jansky's pioneering work did not immediately lead to an expansion in the new science of radio astronomy, but ultimately exploring the radio sky has become one of the most powerful techniques used in understanding the Universe beyond our solar system ◉

COLLISION COURSE

BELOW: The Andromeda Galaxy is shown here in its full glory through an infrared composite image from NASA's Spitzer Space Telescope, which shows the galaxy's older stars (left) and dust (right) separately. Spiral galaxies such as this one tend to form new stars in their dusty, clumpy arms.

Of the six thousand or so stars we can see from Earth with the naked eye, only one object lies beyond the gravitational pull of our galaxy. The picture below is of Andromeda, which is the nearest spiral galaxy to the Milky Way Galaxy and the most distant object visible to anyone who looks up into the night sky with just the naked eye. It may appear as nothing more than a smudge in the heavens, but recent observations by NASA's Spitzer Space Telescope suggest that it is home to a trillion suns.

Andromeda is just one of a hundred billion galaxies in the observable Universe, but there is one thing that singles it out, other than its proximity. While most galaxies are rushing away from each other as the Universe expands, Andromeda is in fact moving directly towards us, getting closer at a rate of around half a million kilometres (310,000 miles) every hour. It seems the two galaxies are destined to meet, guided by the force of gravity.

A galactic collision sounds like a rare and catastrophic event – the meeting of a trillion suns – but in fact such collisions and the resultant mergers of galaxies are not unusual occurrences in the history of the Universe; both the galaxies of Andromeda and the Milky Way have absorbed other galaxies into their structures over the billions of years of their existence.

The sequence of images on the next page has been created as a computer simulation of what would happen during a galactic collision between our neighbour Andromeda and our own Milky Way. The Milky Way Galaxy is shown face-on and you can see it moving from the bottom, up to the left of Andromeda, and then finally to the upper right. From this perspective Andromeda appears tilted.

These images are 1 million light years across, and the timescale between each frame of the sequence is 90 million years. After the initial collision, an open spiral pattern is excited in both the Milky Way and Andromeda, and long tidal tails and the formation of a connecting bridge of stars are apparent. Initially the galaxies move apart one from another, but then they fall back together to meet in a second collision.

As more stars are thrown off in complex ripple patterns, they settle into one huge elliptical galaxy. Spiral galaxies such as Andromeda and the Milky Way are the pinnacle of complexity, order and beauty, but elliptical galaxies are sterile worlds where few stars form. If we humans, and indeed Earth itself, are still here in roughly 3 billion years, this collision will be a spectacular event. Just before we collide, the night sky will be filled by our giant neighbour. When the two galaxies clash there will be so much energy pumped into the system that vast amounts of stars will form, lighting up the whole sky ◉

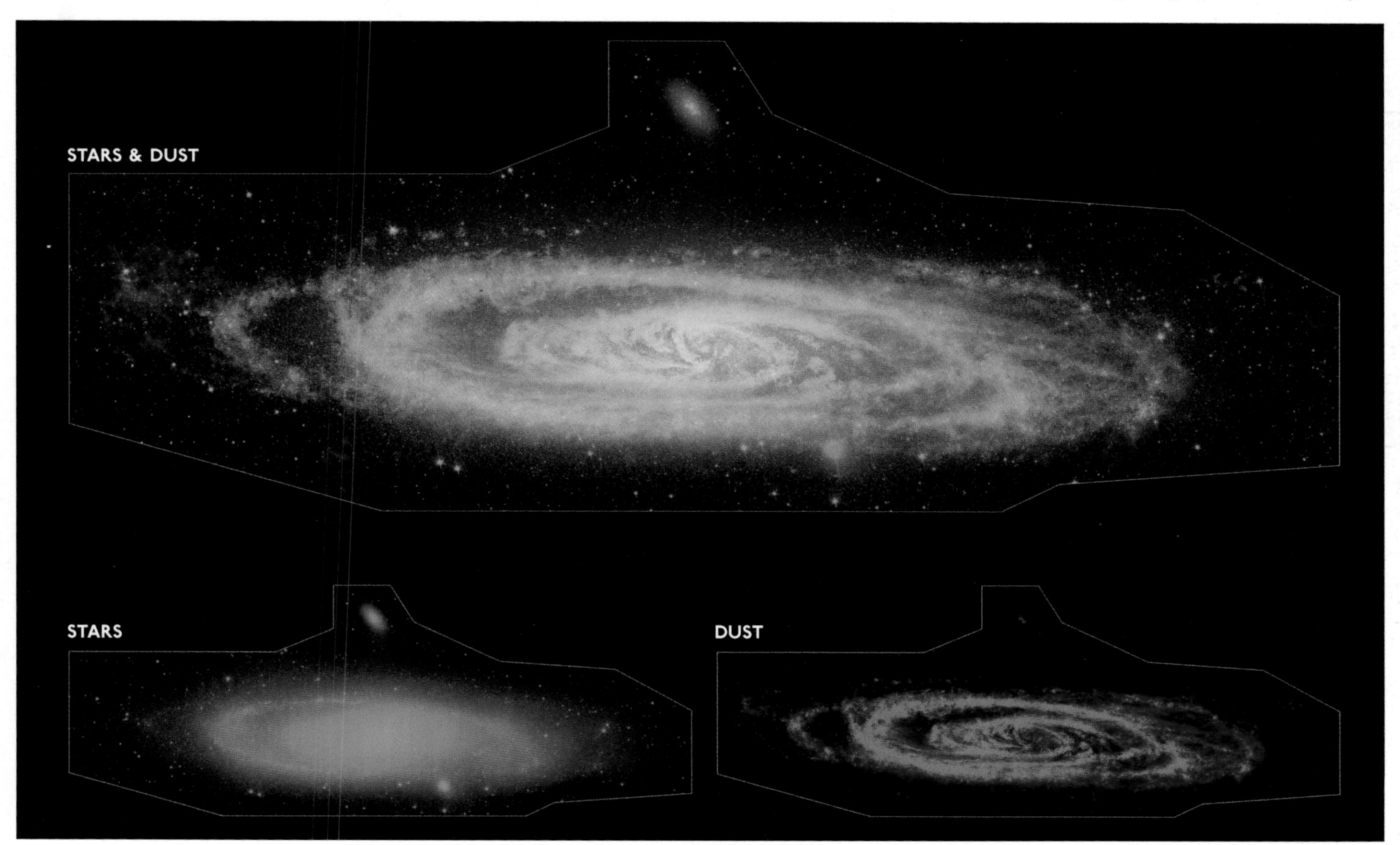

WHEN GALAXIES COLLIDE

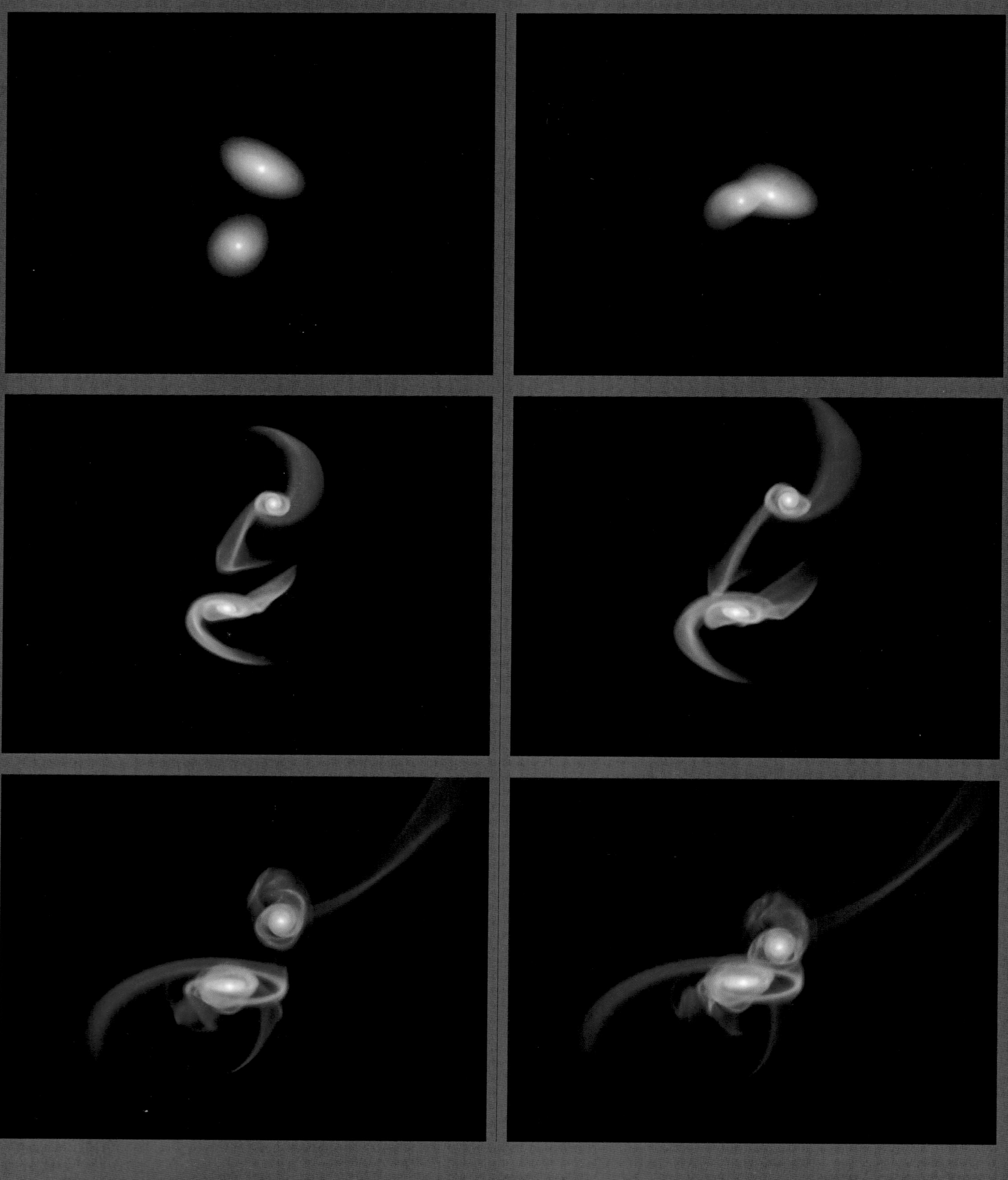

BELOW: This supercomputer animated sequence shows the merger of the Milky Way and Andromeda Galaxies. The sequence begins just before the collision and follows the dynamics of the galaxies until they merge. There are about 90 million years between each frame shown in this sequence.

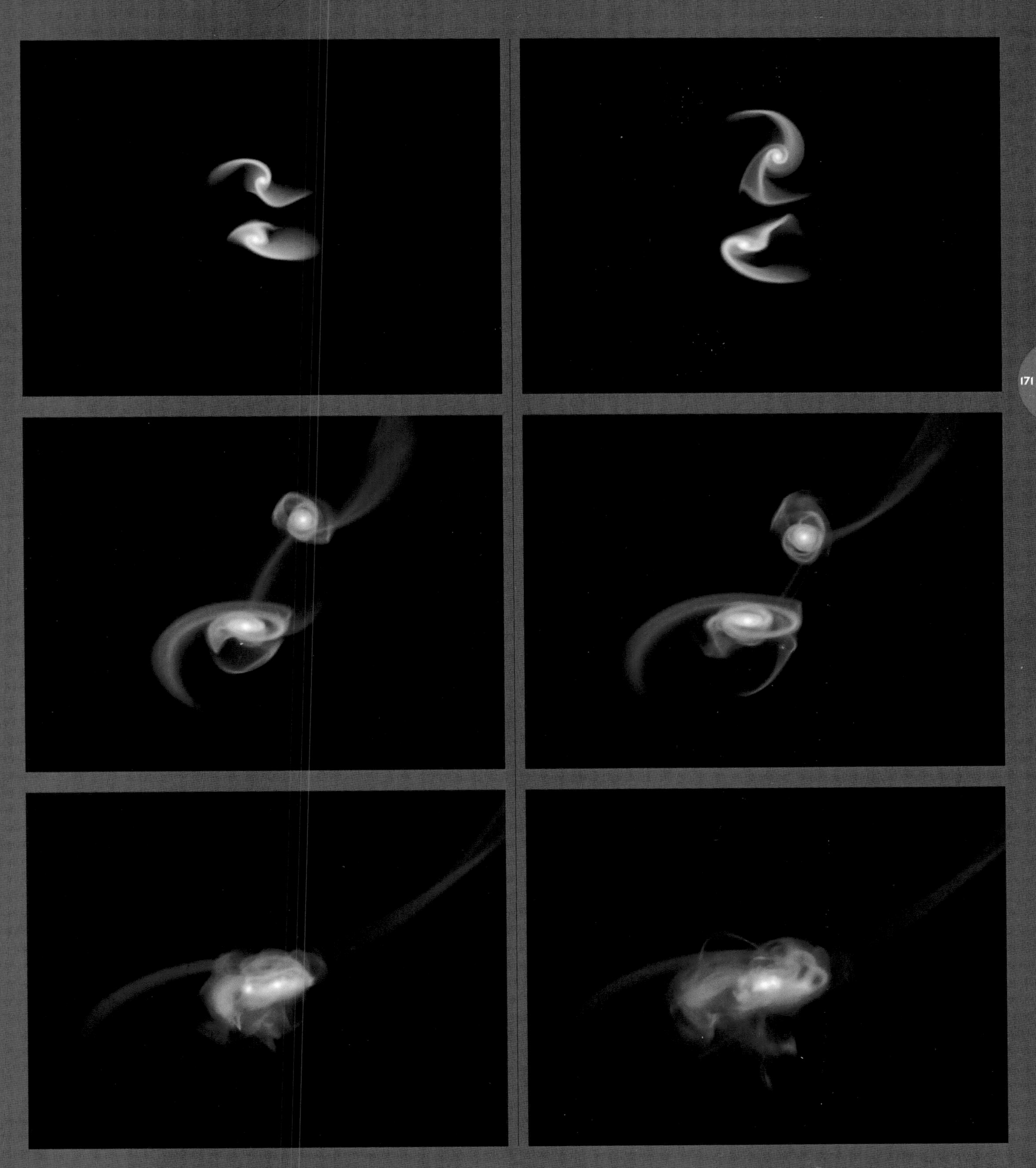

FEELING THE FORCE

Gravity certainly feels like a powerful force. It built our planet, our solar system, and all the billions of star systems in the Universe, diligently assembling clouds of dust and gas into neatly ordered spheres. Matter curves the fabric of the Universe, and in doing so the spheres are bound together and marshalled into orbits, generating the cyclical cosmos we witness from Earth – from our journey through the yearly seasons to the daily ebb and flow of the tides. Gravity reaches far across the space between the star systems, forming galaxies, clusters and superclusters which all beat out orbital rhythms on longer and longer timescales. Gravity is the creator of order and rhythm in our dynamic and turbulent universe.

RIGHT: Galaxy clusters like this one, MS0735.6+7421, are all subject to the power and force of gravity.

THE GRAVITY PARADOX

Despite its reach and influence, there is a mystery surrounding nature's great organisational force; although it is an all-pervasive influence, it is in fact an incredibly weak force – by far the weakest force in the Universe. It is so weak that we overcome it every day in the most mundane of actions. Lift up a teacup and you are resisting the force of gravity exerted on the cup by an entire planet – Earth is trying to stop you, but it is no match for the power of your arm. The reason for this weakness is not known, and the puzzle is brought into stark relief by considering what happens when you lift up the cup. The force that operates your muscles and holds the atoms of your body together is electromagnetism. It is a million million million million million million times stronger than gravity, which is why you will always win in a battle against Earth. Even so, we have evolved to live on the surface of a planet with a particular gravitational field strength, and evolution doesn't produce animals with muscles and skeletons that are stronger than they need to be. Biology rarely wastes precious resources! To demonstrate this, someone at the BBC thought that it would be amusing to see how a human body – mine – would respond if it were transported to a more massive planet.

MY FACE ON A MORE MASSIVE PLANET

The centrifuge at the Royal Netherlands Air Force physiology department was one of the first devices built to spin humans around at speed. Its purpose is to subject fighter pilots to the high G-forces they experience in combat, both for research and to teach them not to black out. As we have discussed, acceleration is indistinguishable from gravity, and spinning around is a good way to achieve high accelerations in a small space. In the case of the human centrifuge, the acceleration is directed towards the centre of the spinning arm, and is caused by the force (known as centripetal force) that acts on your body through the seat to keep you flying in a circle.

My first destination was the gas giant Neptune. Just over seventeen times more massive than Earth, you might expect that the force of gravity would be seventeen times stronger at its surface. However, Neptune's radius is 3.89 times that of Earth at its Equator, so by using Newton's law of gravitation, you'll find that the surface gravity on Neptune is only around 14 per cent greater than Earth's (written as 1.14G). Even with such a small change, I could feel a difference as I lifted up my arms, because they were 14 per cent heavier than normal.

Next up was Jupiter, which is 318 times more massive than Earth. With an equatorial radius 11.2 times greater, the surface gravity would be just over 2.5 times that of our planet. At 2.5G, my arms were 2.5 times heavier than normal, which made them difficult to lift. Apart from this, though, I wasn't in too much discomfort. This all changed when my director decided to send me to exoplanet OGLE2 TR L9b in the constellation of Carina. Over four times the mass of Jupiter, but with a radius only 50 per cent bigger, OGLE2 TR L9b has a surface gravity four times that of Earth. At 4G, things got

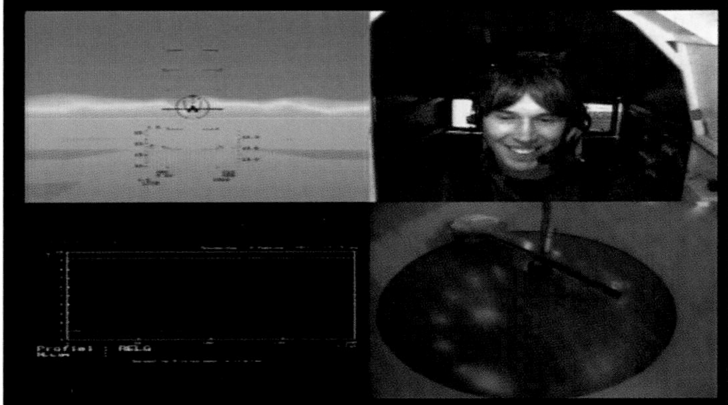

TOP: It may look like a diabolical machine designed to assassinate James Bond and test his escapolgy skills, but this centrifuge at Cologne, Germany, is used to prepare astronauts and fighter pilots for very high G-forces.

quite uncomfortable. I could still speak, but I couldn't lift my arms. It was also quite difficult to breathe because my ribcage and everything else in my body was four times its normal weight, and my muscles aren't used to working that hard.

We then decided to journey beyond OGLE and see how far I could go. As the G-force increased, things got uncomfortable. After a minute or so at 5G, the blood begins to drain from the head, because the heart finds it difficult to pump it up into the brain. This causes faintness and is accompanied by a slight but noticeable narrowing of vision. I had had enough just below 6G, when I was told that my face had been contorted into a funny enough shape to be amusing to the viewers. My job was done. Slowing down was probably more unpleasant than the high-G bit, because the senses are so confused that you feel as if you are tumbling forwards. Gus Grissom described this sensation in the post-flight report of the second manned Mercury mission on Liberty Bell 7, noting that when the main engines shut down after launch, reducing the G-force rapidly, he had to glance at his instruments to reassure himself that his spacecraft was not tumbling.

After my ride I chatted with an F16 pilot who had been subjected to a very fast acceleration and deceleration to 9G. (NATO requires all fighter pilots to be able to deal with this violent ride without passing out.) He told me the centrifuge is far worse than anything you feel in a fighter jet, and having flown in a Lightning and a Hunter, I concur. It's the sustained nature of the G-force in the centrifuge that makes you feel odd; our bodies have not evolved to cope with the weak force of gravity at strengths much greater than those on Earth.

The body with the highest surface gravitational force in the Solar System is the Sun; with a mass 333,000 times that of our planet, it has a surface gravity over 28 times more powerful. The centrifuge cannot go that fast, because this would be a completely unsurvivable G-load.

To find still stronger gravitational fields we have to travel beyond our solar system and look for objects more exotic than mere stars. Our next stop is on one of the strangest worlds in the Universe – one once thought to be populated by aliens ◉

THE LAND OF LITTLE GREEN MEN

In 1967 postgraduate student Jocelyn Bell and her supervisor Anthony Hewish were using a newly completed radio telescope at Cambridge to search for quasars, the most luminous, powerful and energetic objects in the Universe. Quasars, or quasi-stellar radio sources, are now widely believed to be the small, compact regions around supermassive black holes at the centre of very young galaxies. A vast amount of radiation (in excess of the output of an entire galaxy of a trillion suns), is emitted as gas and dust spiral into the black hole.

As Bell and Hewish searched the data for these highly active, ancient galactic centres, they stumbled upon a very strange signal; a pulse that repeated every 1.3373 seconds precisely. It seemed to the Cambridge team to be almost impossible to believe that such a fast regular pulse could come from a natural source, so they named it LGM-1, which stands for Little Green Men.

LEFT: This mosaic image, taken by NASA's Hubble Space Telescope, shows the Crab Nebula, an expanding remnant of a star's supernova explosion. Chinese astronomers recorded this violent event in July 1054, and so too did the people of the Chaco Canyon in New Mexico.

ABOVE: At Chaco Canyon a small, unremarkable-looking painting has been discovered amongst the rocks which probably depicts the explosion of the star that created the Crab Nebula.

If they had discovered a radio beacon from an alien civilisation, you'd have heard about it. The source was entirely natural, as astronomer Sir Fred Hoyle realised immediately on hearing the announcement. However, they had made a new discovery, for which Hewish and fellow astronomer Martin Ryle (though inexplicably and controversially not Bell) received the Nobel Prize in Physics in 1974. Interestingly, though, Bell and Hewish were certainly not the first humans to see one of these wonders – they were beaten to it by an ancient civilisation that witnessed the birth of one almost a thousand years earlier.

CHACO CANYON

A thousand years ago, between AD 900 and 1150, a great civilisation built a series of vast stone structures, known as the Great Houses, along the floor of the arid Chaco Canyon in New Mexico. These buildings remained the largest manmade structures in North America until the nineteenth century. The largest contains more than 700 rooms, many of which are still intact. It is known that these buildings, bizarrely, were not used as permanent residences, because they contain no traces of fires, cooking implements or animal bones. Instead, they seem to have been largely ceremonial; some archaeologists believe that the architecture of the canyon, including its precisely aligned and complex road system, were designed to symbolise and re-enforce the canyon's position not only as the centre of local culture, commerce and religion, but also as the centre of the Universe. The roads and buildings in the canyon and surrounding areas certainly appear to be aligned with the compass points and, it has been suggested, with important moments in the yearly cycle of the Sun – such as the summer and winter solstices. It is difficult to know for sure whether all of the claimed alignments were intentional, but it is known that the Chacoan peoples were keen observers of the skies and possessed a very intricate and advanced cosmology, along with stories of the creation of the constellations and the Universe itself.

One particular site, hidden a mile or so from the main ruins, is the reason for our visit. I have known about it and wanted to come here since I was a little boy; I had no idea where Chaco Canyon was, but I knew about the existence of a small, unremarkable-looking painting on the underside of a rocky overhang next to a dry riverbed half a world away. It was Carl Sagan's *Cosmos*, the book and television series, that introduced me to the wonders of the Universe. In the chapter 'The Lives of the Stars', there is a small black and white photo of the painting, showing three symbols: a handprint, a crescent moon and a bright star. It is known that the painting was made some time around AD 1054, and this was the year of one of the most spectacular astronomical events in recorded history. On 4 July AD 1054, a nearby star exploded. Chinese astronomers recorded the precise date, and the Chacoans would certainly have seen it too because the explosion was visible even in daylight for three weeks, and the fading new star remained visible to the naked eye at night for two years. It would have dominated the skies; a strange and magical sight, perhaps celebrated, perhaps feared; we will never know. We do know precisely where the explosion happened in the sky because its remnant is today one of the most famous and beautiful sights in the heavens: the Crab Nebula.

Apart from the date of the painting, which is not precisely known, the best evidence that this does indeed chronicle the event that the Chinese astronomers recorded is the alignment of the painting. Every 18.5 years, the Moon and Earth will return to the same positions they were in on the nights around 4 July AD 1054. If on one of those rare evenings you go to Chaco Canyon and position yourself beside the painting, the Moon will pass by the position in the sky indicated by the hand print. At that moment, to the left of the Moon, exactly as depicted in the painting, you will see the Crab Nebula.

LEFT: Every 18.5 years, the ruins of the Great Houses of Chaco Canyon and the beautiful rock faces that line the floor of this arid valley are the perfect place from which to see the Crab Nebula in all its glory.

The explosion of 4 July 1054 was a supernova, the violent death of a massive star. It is expected that, on average, there should be around one supernova in our galaxy every century, and this one was almost uncomfortably close, at only 6,000 light years away. The Crab Nebula is the rapidly expanding remains of a star that was once around ten times the mass of our sun; after only a thousand years, the cloud of glowing gas is 11 light years across and expanding at 1,500 kilometres per second. At the heart of the glowing cloud sits the exposed stellar core, which is all that remains of a once-massive sun. It might not look like much when viewed with an optical telescope, but point a radio telescope at it and you will detect a radio signal, pulsing at a rate of precisely 30.2 times a second. It was an object like this that Jocelyn Bell and her colleagues observed in 1967. The Cambridge team weren't listening to little green men, they were listening to the extraordinary sound of a rapidly rotating neutron star – called a pulsar.

Neutron stars are truly amongst the strangest worlds in the Universe; they are matter's last stand against the relentless force of gravity. For most of a star's life, the inward pull of gravity is balanced by the outward pressure caused by the energy released from the nuclear fusion reactions within its core. When the fuel runs out, the star explodes, leaving the core behind. But what prevents this stellar remnant from collapsing further under its own weight? The answer lies not in the physics of stars, but in the world of sub-atomic particles.

The answer to the question of what stops normal matter collapsing in on itself, surprisingly, was not proven until 1967, when physicists Freeman Dyson and Andrew Lenard showed that the stability of matter is down to a quantum mechanical

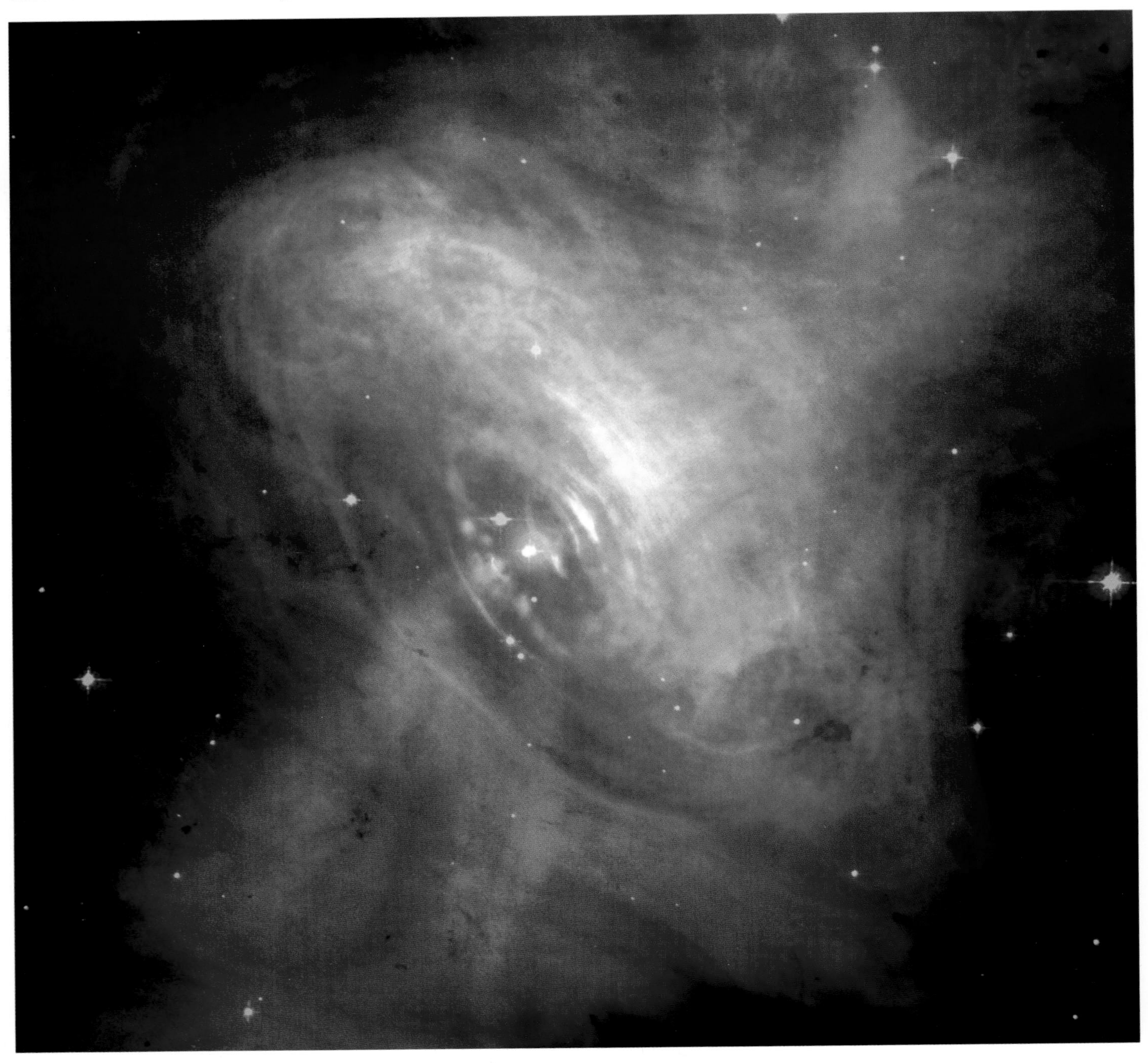

LEFT: Located around 6,000 light-years from Earth, the Crab Nebula is the remnant of a star that exploded as a supernova in AD 1054. This image, taken by NASA's Hubble Space Telescope, shows the centre of the nebula in unprecedented detail.

RIGHT: This composite image of the Crab Nebula has X-ray (blue), and optical (red) images superimposed on it. It is an ever-expanding cloud of gas, and is perhaps the most famous and conspicuous of its kind.

The Crab Nebula is the rapidly expanding remains of a star that was once around ten times the mass of our sun; after only a thousand years, the cloud of glowing gas is 11 light years across and expanding at 1,500 kilometres per second.

effect called the Pauli exclusion principle. There are two types of particles in nature, which are distinguished by a property known as spin. The fundamental matter particles, such as electrons and quarks, and composite particles, such as protons and neutrons, have half-integer spin; these are known collectively as fermions. The fundamental force carrying particles such as photons have integer spin; these are known as bosons. Fermions have the important property that no two of them can occupy the same quantum state. Put more simply, but slightly less accurately, this means you can't pile lots and lots of them into the same place. This is the reason why atoms are stable and chemistry happens. Electrons occupy distinct shells around the atomic nucleus, and as you add more and more electrons, they go into orbits further and further away from the nucleus. It is only the behaviour of the outermost electrons that determine the chemical properties of an element. Without the exclusion principle, all the electrons would crowd into the lowest possible orbit and there would be no complex chemical reactions and therefore no people.

If you try to press atoms together you force their electron clouds together until at some point you are asking all the electrons to occupy the same place (it is more correct to say the same quantum state). This is forbidden, and leads to an effective force that prevents you squashing the atoms together any further. This force is called electron degeneracy pressure, and it is very powerful. In Chapter 4, we will discuss white dwarf stars, the fading embers of suns left to slowly cool after nuclear fusion in their cores ceased. How did they continue to defy the crushing force of gravity? The answer is by electron degeneracy pressure, the dogged reluctance of electrons to being forced too closely together.

But what happens if you keep building more massive white dwarfs, increasing the gravitational force still further? The great Indian astrophysicist Subrahmanyan Chandrasekhar found the answer in one of the landmark calculations of the early years of quantum theory. In 1930, Chandrasekhar showed that electron degeneracy pressure can prevent the collapse of white dwarfs with masses up to 1.38 times the mass

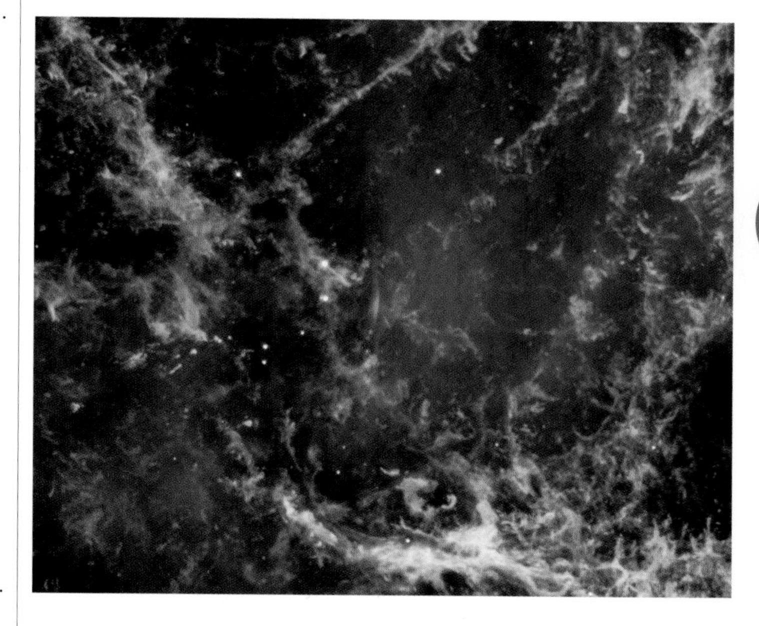

of our sun. For masses greater than this, the electrons won't give in to gravity and move closer together, because they can't. Instead, they give up and disappear.

They don't, of course, vanish into thin air, because they carry properties such as electric charge which cannot be created or destroyed. Instead, the intense force of gravity makes it favourable for them to merge with the protons in the nuclei of the atoms to form neutrons. This is possible through the action of the weak nuclear force in the reverse of the process that turns protons into neutrons in the heart of our sun, allowing hydrogen to fuse into helium. For dying stars with masses above the Chandrasekhar limit, this is the only option, and the entire core turns into a dense ball of neutrons.

Most of the matter that makes up the world around us is empty space. A typical nucleus of a neutron star, which contains virtually all the mass, is around a hundred thousand times smaller in diameter than its atom; the rest is made up of the fizzing clouds of electrons, kept well away from each other by the exclusion principle. If the nucleus were the size of a pea, the atom would be a vast sphere around a hundred metres across, and this is all empty space. With the electrons gone, matter collapses to the density of the nucleus itself; all the space is squashed out of it by gravity, leaving an impossibly

BELOW: This computer simulation of a pulsar shows the beams of radiation emitting from a spinning neutron star. First observed in 1967, the actual mechanism is still the subject of intense theoretical and experimental study.

RIGHT: The Lovell Telescope at Jodrell Bank Centre for Astrophysics aided the exciting discovery of a double pulsar system, announced in January 2004.

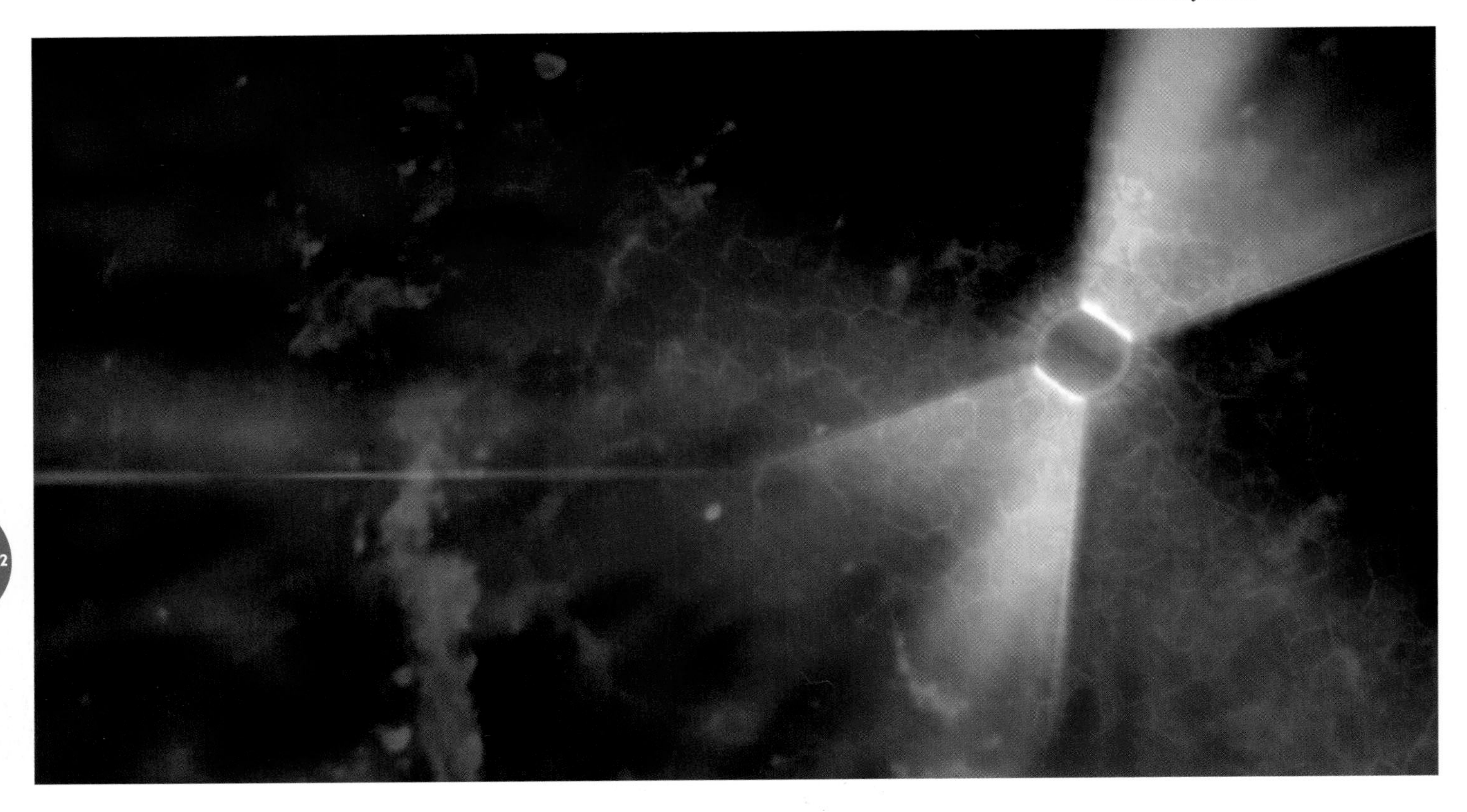

dense nuclear ball. A typical neutron star is around 1.4 times as massive as the Sun, just around the Chandrasekhar limit, crushed into a perfect sphere 20 kilometres (12 miles) across. Neutron star matter is so dense that just one sugar cube of it would weigh more than Mount Everest here on Earth.

The anatomy of neutron stars is still being intensely researched, but they are certainly far more complex than just a ball of neutrons. The surface gravity is of the order of 100,000,000,000G, which is little more than I experienced in the centrifuge. The surface is probably made up of a thin crust of iron and some lighter elements, but the density of neutrons increases as you burrow inwards, for the reasons explained above. Deep in the core, temperatures may be so great that more exotic forms of matter may exist; perhaps quark-gluon plasma, the exotic form of pre-nuclear matter that existed in the Universe a few millionths of a second after the Big Bang.

The unimaginable density and exotic structure aren't the only fantastical feature of neutron stars; many of these worlds, including LGM1 and the neutron star at the heart of the Crab Nebula, have intense magnetic fields and spin very fast. The magnetic field lines, which resemble those of a bar magnet, get dragged around with the stars' rotation, and if the magnetic axis is tilted with respect to the spin axis, this results in two high-energy beams of radiation sweeping around like lighthouse beams. The details of this mechanism are the subject of intense theoretical and experimental study. These are the pulses Bell and Hewitt observed in 1967; the stars are known as pulsars. The fastest known pulsars – millisecond pulsars – rotate over a thousand times every second. Imagine the violence of such a thing; a star the size of a city, a single atomic nucleus, spinning on its axis a thousand times every second.

In January 2004, astronomers using the Lovell Telescope at the Jodrell Bank Centre for Astrophysics, near Manchester, and the Parkes Radio Telescope, in Australia, announced the discovery of a double pulsar system, surely one of the most incredible of all the wonders of the Universe. The system is made up of two pulsars; one with a rotational period of 23 thousandths of a second, the other with a period of 2.8 seconds, orbiting around each other every 2.4 hours. The diameter of the orbit is so small that the whole system would comfortably fit inside our sun. Pulsars are incredibly accurate clocks, allowing astronomers to use the system to test Einstein's theory of gravity in the most extreme conditions known. Imagine the intense warping and bending of space and time close to these two massive, spinning neutron stars. Remarkably, in perhaps the most powerful and beautiful test of any physical theory I know, the predictions of Einstein's Theory of General Relativity, our best current theory of gravity, in the double pulsar system have been confirmed to an accuracy of better than 0.05 per cent. How majestic, how powerful, how wonderful is the human intellect that a man living at the turn of the twentieth century could devise a theory of gravity, inspired by thinking carefully about falling rocks and elevators, that is able to account so precisely for the motion of the most alien objects in the Universe in the most extreme known conditions. That is why I love physics ◉

WHAT IS GRAVITY?

BELOW: Mercury's unpredictable orbit has caused real problems for scientists researching Newton's theory of gravity within the Solar System.

When Newton first published his Law of Universal Gravitation in 1687 he transformed our understanding of the Universe. As we have seen, his simple mathematical formula is able to describe with unerring precision the motion of moons around planets, planets around the Sun, solar systems around galaxies, and galaxies around galaxies. Newton's law is, however, only a model of gravity; it has nothing at all to say about how gravity actually is, and it certainly has nothing to say about a central mystery: why do all objects fall at the same rate in gravitational fields? This question can be posed in a different way by looking again at Newton's famous equation:

$$F = G\frac{m_1 m_2}{r^2}$$

This states that the gravitational force between two objects is proportional to the product of their masses – let's say that m_1 is the mass of Earth and m_2 is the mass of a stone falling towards Earth. Now look at another of Newton's equations: $F = ma$, which can be written with a bit of mathematical rearrangement as $a = F/m$. This is Newton's Second Law of Motion, which describes how the stone accelerates if a force is applied to it. It says that the acceleration (a) of the stone is equal to the force you

WHY DO OBJECTS FALL AT THE SAME RATE?

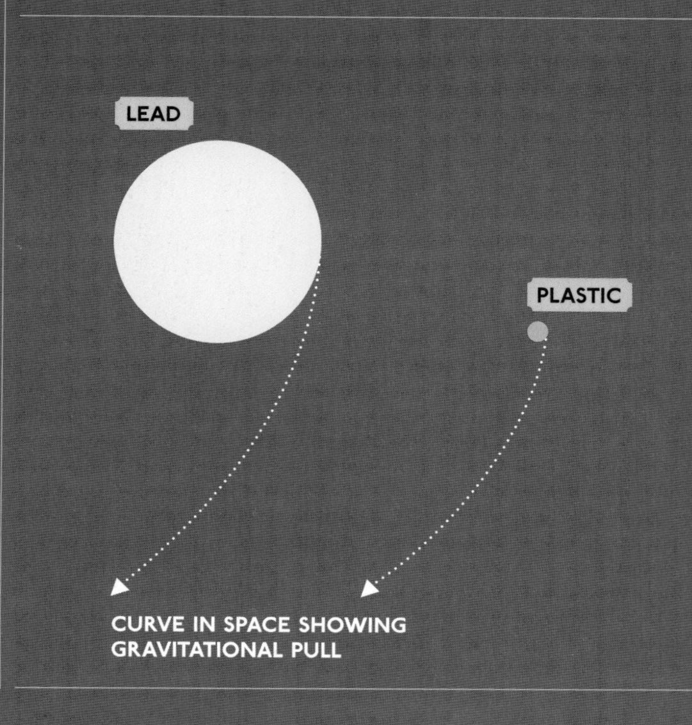

apply to it (F) divided by its mass (m). The reason why things fall at the same rate in a gravitational field, irrespective of their mass, is that the mass of the stone in these two equations (labelled m_2 in the first equation and m in the second), are equal to each other. This means that when you work out the acceleration, the mass of the stone cancels out and you get an answer which only depends on the mass of Earth – the famous 9.81 m/s². We said this earlier in the chapter in words: if you double the mass of something falling towards Earth, the gravitational force on it doubles, but so does the force needed to accelerate it. But there is a very important assumption here that has no justification at all, other than the fact that it works: why should these two masses be the same? Why should the so-called inertial mass – which appears in $F = ma$ and tells you how difficult it is to accelerate something – have anything to do with the gravitational mass, which tells you how gravity acts on something? This is a very important question, and Newton had no answer to it.

Newton, then, provided a beautiful model for calculating how things move around under the action of the force of gravity, without actually saying what gravity is. He knew this, of course, and he famously said that gravity is the work of God.

If a theory is able to account for every piece of observational evidence, however, it is very difficult to work out how to replace it with a better one. This didn't stop Albert Einstein, who thought very deeply about the equivalence of gravitational and inertial mass and the related equivalence between acceleration and the force of gravity. At the turn of the twentieth century, following his great success with the Special Theory of Relativity in 1905 (which included his famous equation $E=mc^2$), Einstein began to search for a new theory of gravitation that might offer a deeper explanation for these profoundly interesting assumptions.

Although not specifically motivated by it, Einstein would certainly have known that there were problems with Newton's theory, beyond the philosophical. The most unsettling of these was the distinctly problematic behaviour of a ball of rock that was located over 77 million kilometres (48 million miles) from Earth.

The planet Mercury has been a source of fascination for thousands of years. It is the nearest planet to the Sun and is tortured by the most extreme temperature variations in the Solar System. Due to its proximity to our star, Mercury is a difficult planet to observe from Earth, but occasionally the planets align such that Mercury passes directly across the face of the Sun as seen from Earth. These transits of Mercury are one of the great astronomical spectacles, occurring only 13 or 14 times every century. Mercury has the most eccentric orbit of any planet in the Solar System. At its closest, Mercury passes just 46 million kilometres (28 million miles) from the Sun; at its most distant it is over 69 million kilometres (42 million miles) away. This highly elliptical orbit means that the speed of the movement of this planet varies a lot during its orbit, which means in turn that very high-precision measurements were necessary to map its orbit and make predictions of its future transits. Throughout the seventeenth and eighteenth centuries, scientists would gather across the globe to watch the rare transits of Mercury. These scientists used Newton's Law of Gravity to predict exactly when and where they could view the spectacle, but it became a source of scientific fascination and no little embarrassment when, time after time, Mercury didn't appear on cue. The planet regularly crossed the Sun's disc later than expected, sometimes by as much as several hours.

Mercury's unusual orbit was a real problem, but because of the observational uncertainties it wasn't until 1859 that the French astronomer Urbain Le Verrier proved that the details of Mercury's orbit could not be completely explained by Newtonian gravity. To solve the problem, many astronomers reasoned there must be another planet orbiting between the Sun and Mercury. This planet had to be invisible to our telescopes, but it must also exert a gravitational force large enough to disturb Mercury's orbit. Encouraged by the recent discovery of the planet Neptune, based on a similar anomaly in the orbit of Uranus, they named the ghost planet Vulcan ◉

MERCURY'S ORBIT AROUND THE SUN

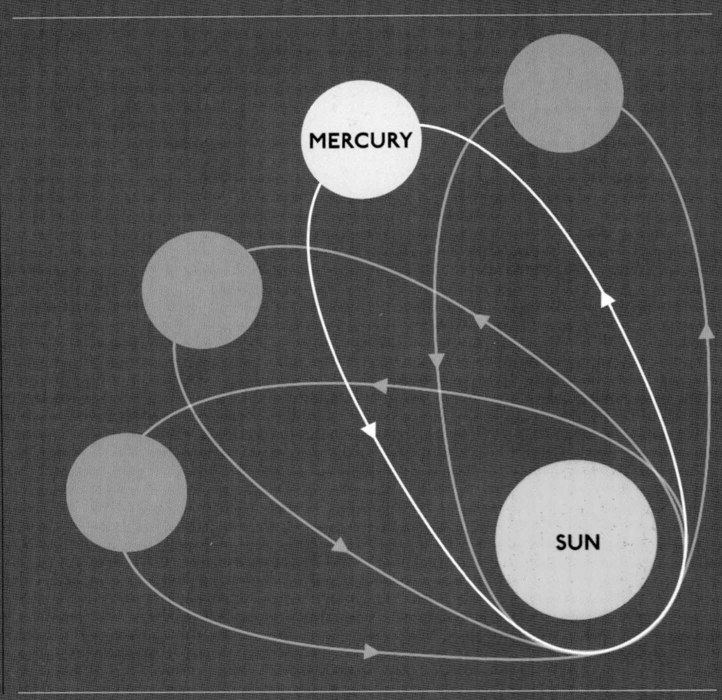

For decades astronomers searched and searched for Vulcan, but they never found it. The reason for this is that Vulcan doesn't exist. The errors in the predictions in fact signalled something far more profound: Newton's Theory of Universal Gravitation is not correct.

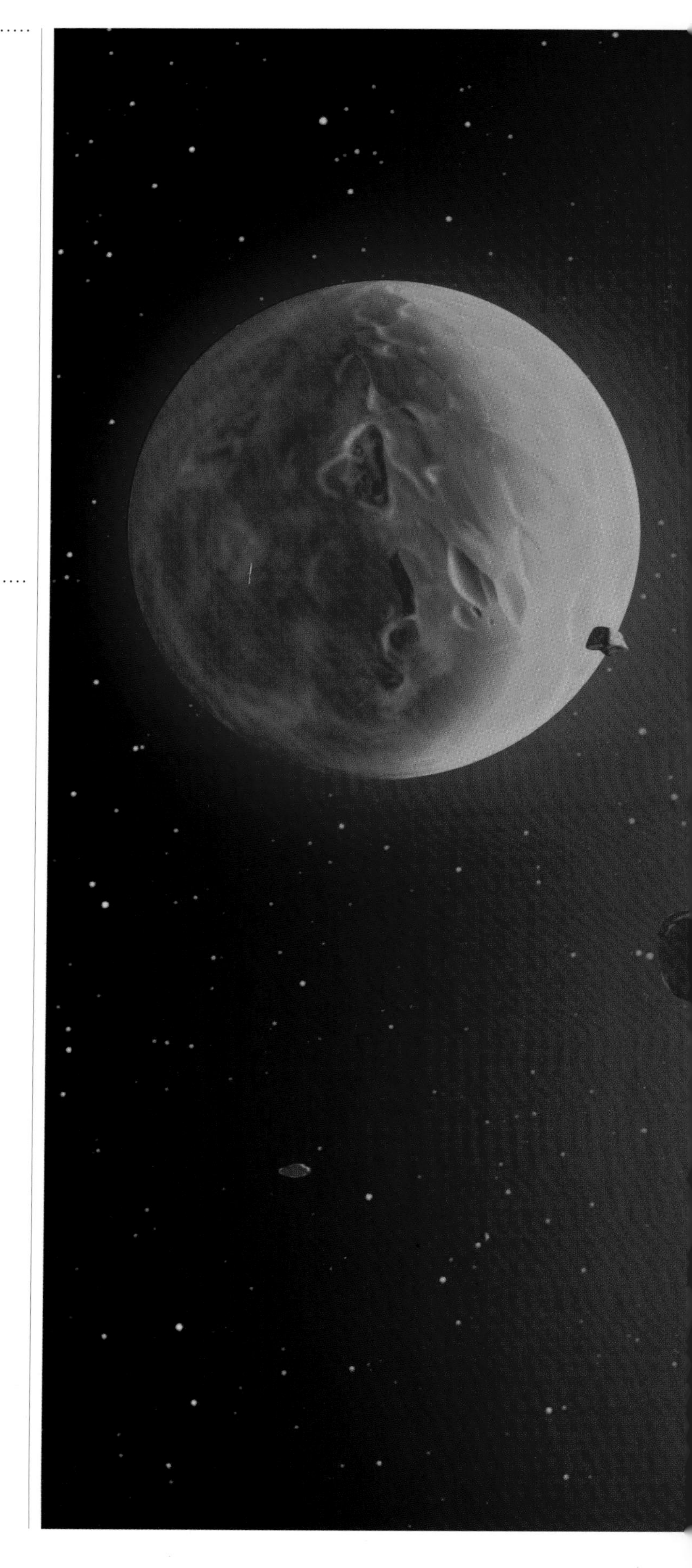

RIGHT: This image shows an artist's impression of the hypothetical planet Vulcan, which was once believed to orbit in an asteroid belt closer to the Sun than Mercury. Vulcan was supposedly first sighted by amateur astronomer Lescarbault on 26 March 1859, but further observations were inconclusive and Vulcan was later proved to be a ghost planet.

EINSTEIN'S THEORY OF GENERAL RELATIVITY

LEFT: German-born physicist Albert Einstein (left) created the famous Theory of General Relativity. British astrophysicist Sir Arthur Eddington (right), later put this theory to the test and confirmed its accuracy.

Einstein would have loved the Vomit Comet. The fact that the effects of gravity can be completely removed by falling freely in a gravitational field was, for him, the thought experiment that led to his theory of General Relativity. How wonderful it would have been for him to experience it as I did! The reason I say this is that, as I floated next to my little plastic Albert in the Vomit Comet, I understood very deeply why Einstein was so interested in freefall. The point is this; inside the plane, falling towards Earth, it is absolutely impossible to tell that you are moving. It is impossible to tell that you are near a planet. It is impossible to tell that, according to someone stood on the ground, you are accelerating at 9.81 m/s^2 towards the ground. You are simply floating, along with everything else in the plane. I let some little drops of water out of a bottle and they floated in front of my face; the cameraman and director floated next to the water droplets, little plastic Albert and me. There was self-evidently no force acting on anything at all, otherwise things would have moved around.

And yet, from the point of view of someone on the ground, we were flying in a parabolic arc, moving forwards through the air at hundreds of miles an hour and accelerating violently towards the ground. The force of gravity is very much present in this description. Einstein's theory takes the view that the two ways of looking at the Vomit Comet – from inside and outside – should be treated as equivalent. No one inside the plane or out has the right to claim that they are right and the other is wrong! If, inside the plane, there is no experiment you can do to prove that you are accelerating towards the ground, you are well within your rights to claim that you are not. Acceleration has cancelled out gravity. Of course, you could look out of the windows, but even then you could claim that Earth is accelerating up towards you and that you are simply floating. From this perspective, everyone on Earth feels a gravitational force pulling them onto the ground because they are being accelerated upwards at a rate of 9.81 m/s^2. Acceleration is therefore equivalent to gravity; this is known as the equivalence principle, and it was very important to Einstein.

So what, then, is gravity? The explanation in Einstein's theory is beautifully simple: gravity is the curvature of spacetime.

In technical language, Einstein would have defined the Vomit Comet, during its time in freefall, as an inertial frame of reference – which is to say that it can be legitimately considered to be at rest, with no forces acting on it.

The assertion that sitting in a falling aircraft should be considered as being absolutely equivalent to floating around in space, far beyond the gravitational pull of any planet or moon, can be used to explain why all objects fall at the same rate.

Why? Simply because there are two equally valid ways of looking at what is happening. From the point of view inside the plane, nothing at all is happening; everything is simply floating, untouched by any forces of any kind. If no forces are acting, then everything naturally stays where it is put. Shift outside the plane, however, and things appear different; everything is falling towards Earth, accelerating under the action of the force of gravity. But, very importantly, the reality of the situation cannot change depending on which point of view we adopt – everything has to behave in the same way in reality, irrespective of how you look at it. If plastic Albert and the globules of water float in front of my face when viewed from my vantage point inside the plane, then plastic Albert and the globules of water had better float in front of my face when viewed from a vantage point outside the plane. In other words, we had all better be accelerating towards the ground at exactly the same rate! Notice that we've made no assumptions about the equivalence between gravitational and inertial masses here; we've just said that a freely falling box in Earth's gravitational field is indistinguishable from a freely falling box in space, or indeed any freely falling box anywhere in the Universe, around any planet, any star, or any moon.

So what, then, is gravity? The explanation in Einstein's theory is beautifully simple: gravity is the curvature of spacetime. What is spacetime? Spacetime is the fabric of the Universe itself.

A good way to picture spacetime, and what it means to curve it, is to think about a simpler surface; the surface of Earth. Our planet has a two-dimensional surface, which is to say that you only need two numbers to identify any point on it: latitude and longitude. Earth's surface is curved into a sphere, but you don't need to know that to move around on it and navigate from place to place. The reason we can picture the curvature is that we are happy to think in three dimensions, so we can actually see that Earth's surface is curved. But imagine that we were two-dimensional beings, confined to move on the surface of Earth with absolutely no concept of a third dimension. We would know nothing about up and down, only about latitude and longitude. It would be very difficult indeed for us to picture in our mind's eye the curvature of our planet's surface.

RIGHT: As light has no mass, Newton's theory states that it is not affected by gravity, although observation does infact show that it is.

Now let's extend our analogy to see how the curvature of something can give rise to a force. Imagine that a pair of two-dimensional friends are standing on the Equator and decide to take a journey due north. They decide to walk parallel to each other, with the intention of never bumping into each other. If they both keep walking, they will walk up parallel lines of longitude, and they will find that as they get closer and closer to the North Pole they will get closer and closer together. Eventually, when they reach the North Pole, they will bump into each other! As three-dimensional beings, we can see what happened; Earth's surface is curved, so all the lines of longitude meet at the poles. However, from the perspective of our two-dimensional friends, even though they kept assiduously to their parallel lines they still were mysteriously drawn together. They may well conclude from this that a force was acting between them, attracting them towards one another. In Einstein's theory, that force is gravity.

The complicated bit about Einstein's Theory of General Relativity is that the surface we need to think about, spacetime, is not two-dimensional but four-dimensional. It is a mixture of the familiar three dimensions of space, plus an additional dimension of time mixed in. It will take us too far from our story to explore spacetime in detail, but it was found to be necessary by Einstein and others at the turn of the twentieth century to explain, in particular, the behaviour of light and the form of Maxwell's equations that we met in Chapter 1. Suffice to say that the surface of our universe, on which we all live our lives, is four-dimensional. What Einstein showed is that the presence of matter and energy – in the form of stars, planets and moons– curves the surface of spacetime, distorting it into hills and valleys. His equations describe exactly what shape spacetime should be around any particular object, such as the Sun, for example, and they also describe how things move over the curved surface. And here is the key point: just like our two-dimensional friends, things move in straight lines; but just like our two-dimensional friends, this isn't what it looks like if you don't know that spacetime is curved. When you're moving across the curved surface, it appears that a force is acting on you, distorting your path. One of the first things Einstein did with his new, geometric theory of gravity was to calculate what Mercury's straight-line path through the curved spacetime around the Sun would look like to us, trapped on the surface of spacetime. To his delight, he found that Mercury would orbit the Sun, and in precisely the way that had been observed over the centuries of transit observations. Where Newton failed, Einstein succeeded.

Einstein had found a completely geometrical way of describing the force of gravity, and it is quite wonderfully elegant. Not only does it predict the orbit of Mercury, but it also provides a very appealing explanation for the equivalence principle. Why do all objects fall at the same rate in a gravitational field, irrespective of their mass or composition? Because the path they take has nothing to do with them at all – they are simply following straight-line paths through the curved spacetime.

Perhaps the most startling demonstration of this is the bending of light by gravity. Light has no mass, and so in Newton's theory it shouldn't be affected by gravity at all. However, according to Einstein's theory, it doesn't matter that it has no mass, it will still be following a straight line through the curved spacetime, so it will appear to follow exactly the same path as everything else. Let's do a thought experiment to see how strange this is. Stand on the ground (on a very, very big planet – I'll explain why I said this in a moment!) with a rock in one hand and a laser beam in the other. Point the laser beam horizontally, drop the rock and fire the laser. Which one hits the ground first? The answer is that they both hit the ground at the same time, because they both move through the same curved space. Light falls at the same rate in a gravitational field as everything else. Now, there is a caveat here. Why did I say a very very big planet? Because light travels at almost 300,000 kilometres per second, so if the rock takes a second to hit the ground, so will the light. But it will

THE GRAVITY CONUNDRUM

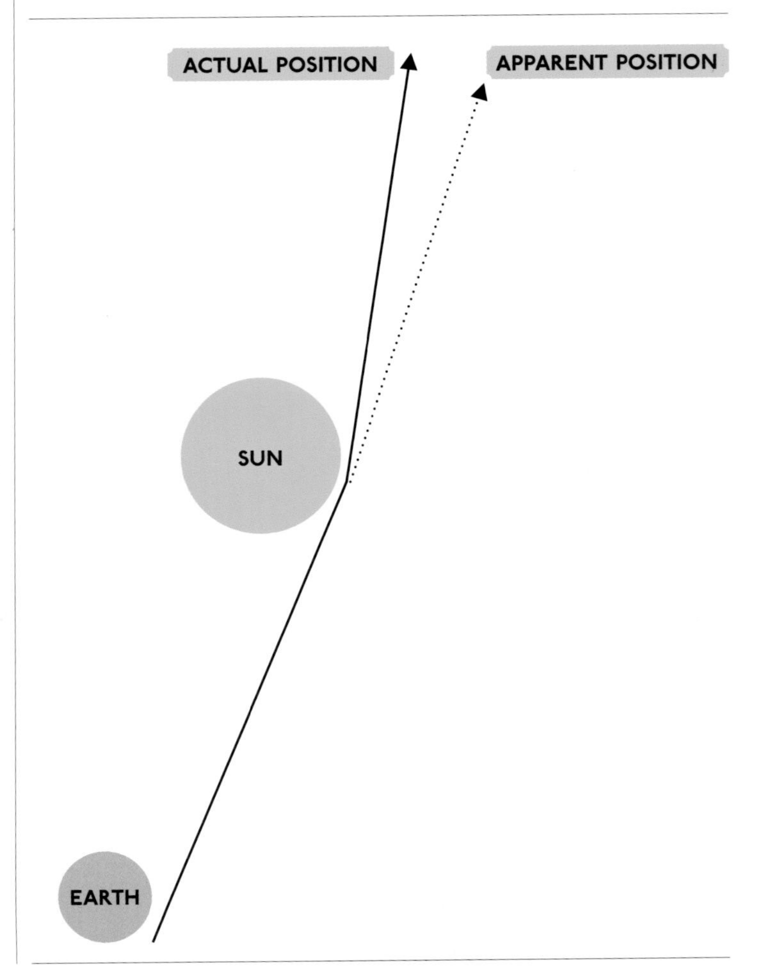

have flown 300,000 kilometres in the horizontal direction by the time it reaches the ground, and on Earth that would mean the surface of the planet had long since curved away! However, the principle still holds.

As an interesting aside, what would happen if you fired the laser beam directly at the ground? Light must always travel at the same speed, it can't speed up, so it will travel towards the ground at exactly 299,792,458 metres per second. But shouldn't it accelerate at 9.81 m/s^2 as it drops? No, it can't, because it always travels at exactly 299,792,458 metres per second. So what happens? Well, the energy of the light can change, although the speed cannot, so the light gets shifted towards the blue end of the spectrum as it flies towards the ground and gains energy from its fall. That is to say that its wavelength gets shorter and its frequency increases. This is very interesting because the second is defined as the length of time it takes a fixed number of wavelengths of a particular colour of light to pass by an observer. Let's say that you use the frequency of the laser beam held in your hand to synchronize a clock, then you fire the laser at the ground; when the light hits the ground, its frequency will have increased. This means that the peaks and troughs of the laser light beam are arriving more frequently than they did when they set off. So, from the point of view of someone on the ground, the clock above the ground will be running

In the language of General Relativity, we might say that the presence of Earth bends spacetime near it such that time passes more slowly than it does far away.

slightly fast. Is this true? Yes, it is. The effect is known as gravitational time dilation; gravity slows down time, so clocks close to the ground run slower than those in orbit. In the language of General Relativity, we might say that the presence of Earth bends spacetime near it such that time passes more slowly than it does far away. This is a very real effect and is one that has to be taken into account in the GPS satellite navigation system, which relies on precise timekeeping to measure distances. The GPS satellites orbit at an altitude of 20,000 kilometres (12,500 miles), which means that their clocks run faster than they do on the ground by 45 microseconds per day, because they are in a weaker gravitational field. The fact that they are moving relative to the ground also affects the rate of their clocks, and when everything is taken into account the timeshift reduces to 38 microseconds per day. This would be equivalent to a distance error of over 10 kilometres (6 miles) per day, which would make the system useless. So, every time we get into our cars and use satellite navigation, we are using Einstein's theory of gravity in order to correctly ascertain our position on the surface of Earth.

To summarise, then, had Einstein experienced the Vomit Comet, he would have described it, during its time in freefall, as following a straight-line path through spacetime. As long as it continues on this path, the plane and its passengers will not feel the force of gravity at all; it is only when something stops the plane following its straight-line path through spacetime that a force is felt. If the plane didn't stop itself falling, this obstacle would be the ground!

It is worth making a final brief aside here, which also serves to underline what we've just learnt. The experimental fact that triggered all this discussion is that the gravitational and inertial masses of objects are the same. Einstein provides a natural explanation for this: gravity is simply a result of the fact that there is such a thing as spacetime, and that it is curved, and that things move in straight lines through this curved spacetime. It is also possible to take a different view; there could be some deep reason why the gravitational and inertial masses of things are equal – a reason that we have yet to discover. The fact that they are equal allows us to build a geometric theory of gravity. In that case, Einstein's theory might more properly be considered to be a model, in the same way that Newton's theory is a model. At the moment we have no way of deciding between these two possibilities, but it's worth being aware that they are both valid ways of looking at the situation.

Einstein's Theory of General Relativity is rightly considered to be one of the great intellectual achievements of all time. It is conceptually elegant and probably the theory that physicists most often attach the word 'beautiful' to. Ultimately, though, it doesn't matter how beautiful a theory is, the only thing that matters is that its predictions are in accord with our observations of the natural world. The orbit of Mercury is one such observation; the slowing down of time in gravitational fields is another; but to really test Einstein's theory to the limit, we have to journey far out into space and visit the most exotic and massive objects in the known Universe – places where the force of gravity becomes exceptionally strong ◉

INTO THE DARKNESS

The success of Einstein's Theory of General Relativity is one of the greatest of human achievements, and in my view it will be remembered as such for as long as there is anything worth calling a civilisation. But there is a final twist to the story of gravitation, because Einstein's remarkable theory predicts its own demise.

The collapse of a neutron star is prevented by neutron degeneracy pressure. Neutrons are fermions, as are electrons, but because they are more massive than electrons, they can be packed much more tightly together before the Pauli exclusion principle steps in once more and forbids further contraction. Another stable staging post against gravity should be provided by quark degeneracy pressure, because quarks too are fermions, but ultimately, if the star is too massive gravity will overwhelm even these fantastically dense objects. It is believed that the limit above which no known law of physics can intervene to stop gravity is around three times the mass of the Sun. This is known as the Tolman-Oppenheimer-Volkoff limit. For the remnants of stars with masses beyond this limit, gravity will win.

In 1915, only one month after Einstein published the Theory of General Relativity, the physicist Karl Schwarzschild found a solution to Einstein's equations which is now known as the Schwarzschild metric. The Schwarzschild metric describes the structure of spacetime around a perfectly spherical object. There are two interesting features of Schwarzschild's spacetime: one occurs at a particular distance from the object, known as the Schwarzschild radius, but for distances less than the Schwarzschild radius, space and time are distorted in such a way that the entire future of anything that falls in will point inwards. This sounds weird, but remember that space and time are mixed up together in Einstein's theory. In more technical language, we say that the future light cones inside the Schwarzschild radius all point towards the centre. This means that, as inexorably as we here on Earth march into the future, if you were to cross the line defined by the Schwarzschild radius, you would inexorably march inwards towards the object that is bending spacetime. There would be no escape, not even for light itself, in the same way that you cannot escape your future. This surface, defined by the Schwarzschild radius surrounding the object, is known as the event horizon. But what has happened to the object itself? This is the second interesting feature of the Schwarzschild metric. Let's first think about the Sun. If you asked what the Schwarszchild radius for a star with the mass of the Sun is, it would be 3 kilometres (1 mile). This is inside the Sun! So there is no problem here, because you can't get that close to the Sun without actually being inside it, at which point all the mass outside you doesn't count any more.

But what about an object like a collapsing neutron star, getting smaller and smaller and denser and denser? What if you could have an object that was dense enough to have the mass of the Sun and yet be physically smaller than the Schwarszchild radius? It seems that there are such objects in the Universe; the stars for which even neutron degeneracy

Black holes are fascinating objects; we don't understand them, and yet we know they exist. They are of immense importance ... the physics that lies inside the event horizon is undoubtedly fundamental.

pressure will not suffice to resist the force of gravity. These objects are called black holes. At the very centre of the black hole, at *r=0*, the Schwarzschild metric has another surprise in store; the spacetime curvature becomes infinite. In other words, the gravitational field becomes infinite. This is known as a singularity. In physical theories, the existence of singularities signals the edge of the applicability of the theory; in simple language, there must be more to it! This has led many physicists to search for a new theory of gravity. Quantum theories of gravity such as string theory may be able to avoid the appearance of singularities, by effectively setting a minimum distance scale below which spacetime does not behave in the manner described by Einstein's equations.

As yet, we do not know whether any of these current theories are correct, or even if they are on the road to being correct, but what we do know is that black holes exist. At the centre of our galaxy, and possibly every galaxy in the

LEFT: This coloured X-ray image shows the area around the supermassive black hole, known as Sagittarius A*, which sits at the centre of the Milky Way Galaxy.

BELOW: This artist's impression helps us to visualise the mysterious objects in space that are black holes.

Universe, there is believed to be a supermassive black hole. Astronomers believe this because of precise measurements of the orbit of a star known as S2. This star orbits around the intense source of radio waves known as Sagittarius A* that sits at the galactic centre. S2's orbital period is just over 15 years, which makes it the fastest-known orbiting object, reaching speeds of up to 2 per cent of the speed of light. If the precise orbital path of an object is known, the mass of the thing it is orbiting can be calculated, and the mass of Sagittarius A* is enormous – 4.1 million times the mass of our Sun. Since the star S2 has a closest approach to the object of only 17 light hours, it is known that Saggitarus A* must be smaller than this, otherwise S2 would literally bump into it. The only known way of cramming 4.1 million times the mass of the Sun into a space less than 17 light hours across is as a black hole, which is why astronomers are so confident that a giant black hole sits at the centre of the Milky Way. These observations have recently been confirmed and refined by studying a further 27 stars, known as the S-stars, all with orbits taking them very close to Sagittarius A*.

Black holes are fascinating objects; we don't understand them, and yet we know they exist. They are of immense importance, because despite the fact that we will never encounter one directly, the physics that lies inside the event horizon is undoubtedly fundamental. These are objects that will require a new theory of gravity, indeed a new theory of space and time, to describe. One of the holy grails of observational astronomy is to find a pulsar orbiting around a black hole. Such a system surely exists somewhere, and to be able to observe the behaviour of one of these massive cosmic clocks in the intensely curved spacetime close to a black hole would surely test Einstein's Theory of General Relativity to its limit. It may even, if we are lucky, reveal flaws that point us towards a new theory ◉

THE ANATOMY OF A BLACK HOLE

For all their mystery, we do know that black holes exist. The idea of a body so massive that even light could not escape its grip was first suggested in the eighteenth century, and today we now know that there is not only a black hole at the centre of our galaxy, but also possibly in the centre of every galaxy. We may never directly see one, but the secrets they contain may one day help us answer some of the most fundamental questions in the Universe.

Immense gravitational pull from the mass of the black hole

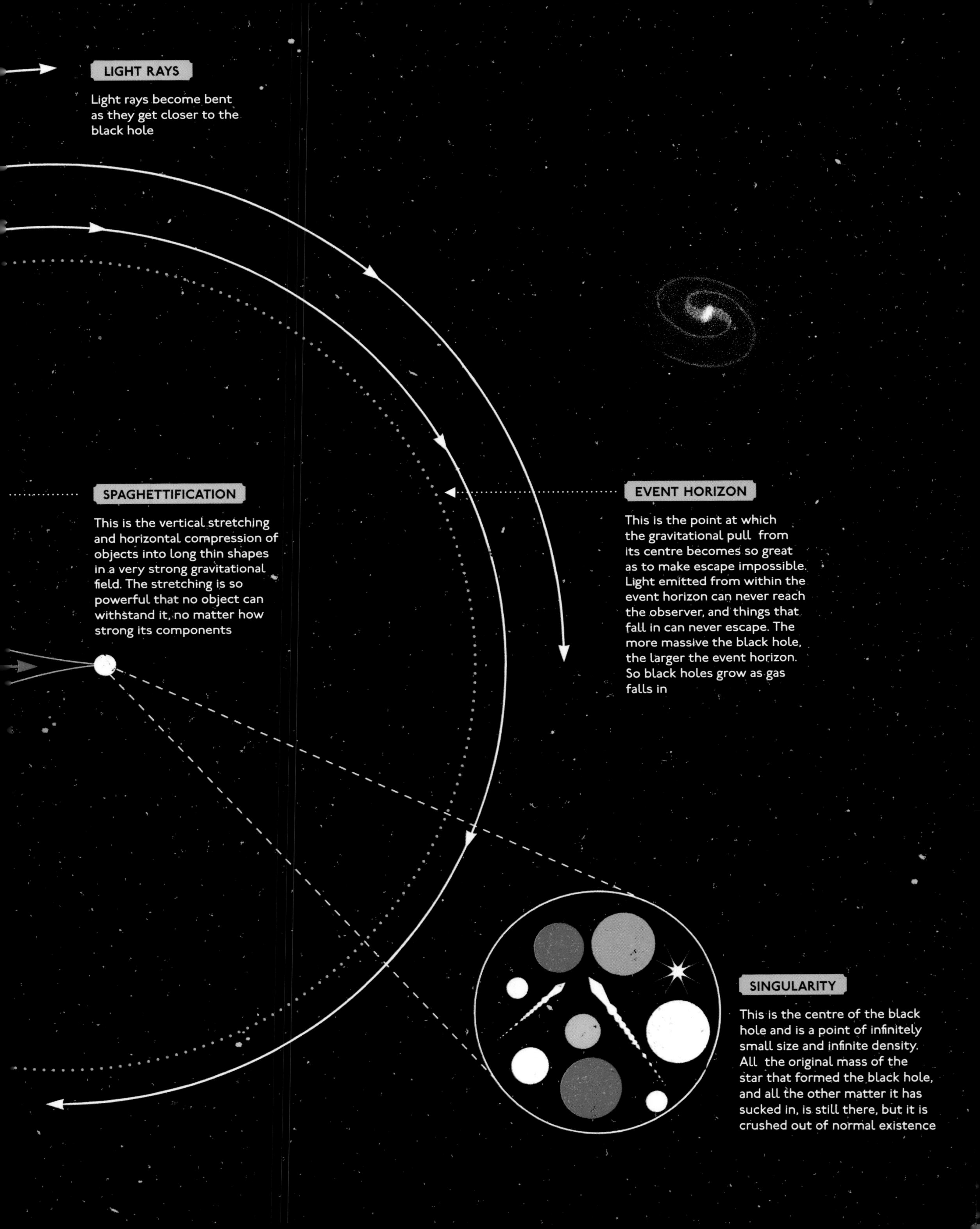

LIGHT RAYS

Light rays become bent as they get closer to the black hole

SPAGHETTIFICATION

This is the vertical stretching and horizontal compression of objects into long thin shapes in a very strong gravitational field. The stretching is so powerful that no object can withstand it, no matter how strong its components

EVENT HORIZON

This is the point at which the gravitational pull from its centre becomes so great as to make escape impossible. Light emitted from within the event horizon can never reach the observer, and things that fall in can never escape. The more massive the black hole, the larger the event horizon. So black holes grow as gas falls in

SINGULARITY

This is the centre of the black hole and is a point of infinitely small size and infinite density. All the original mass of the star that formed the black hole, and all the other matter it has sucked in, is still there, but it is crushed out of normal existence

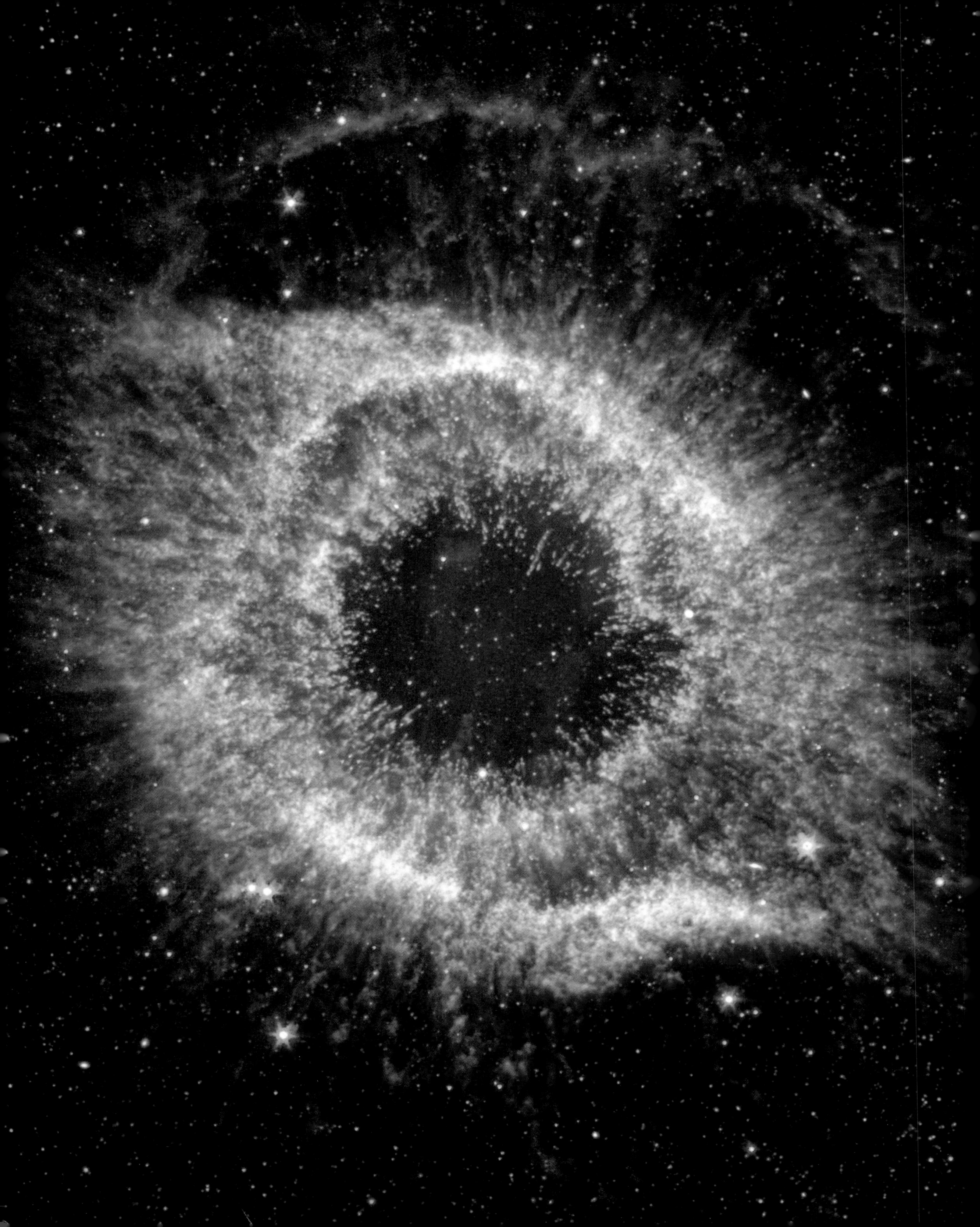

CHAPTER 4

DESTINY

THE PASSAGE OF TIME

This is the story of something so fundamental that it's impossible to imagine a universe without it, yet it is a property of the Universe that modern science still struggles to explain. Time is something that feels very human; it regulates our days and its relentless and unavoidable passing drives our lives forward. It is why each one of us has a beginning and an end. But time isn't a human creation; we evolve with its passing, but so does the rest of the Universe. Time is woven into the very fabric of the cosmos. Even with our incomplete understanding, our exploration of time has allowed us to do something remarkable: just by investigating the nature of time and the natural world as we find it here on Earth, we've been able to not only glimpse the beginning of the Universe, but to imagine how it might end.

BELOW: The towers of the ruined temple of the ancient hilltop fortress at Chankillo are a remarkable sight, standing tall through the sand-laden skies of the Peruvian desert.

OVERLEAF: The Thirteen Towers of the temple at Chankillo are believed to serve a dual purpose: they are also an ancient calendar. The towers are carefully placed to use sunrise to mark the passing of the days.

On the arid coastal plain of northwestern Peru lies one of South America's greatest astronomical secrets. Few people know about the hilltop fortress at Chankillo, and even fewer visit it, but for archaeologist and astronomer alike it is both evocative and fascinating. Two and a half thousand years ago, a civilisation we know almost nothing about built a city in this inhospitable place. The grandest of the structures was a fortified temple with walls of brilliant white covered with red-painted figures. Commanding a sweeping view across the desert, the temple would have dominated the sand-laden skies, however, today all but the smallest fragments of the decorations are gone, dulled by passing centuries. The building's location has puzzled archaeologists for many years because, while it is commanding, the hilltop site is not the best defensive position in the area, and it is unimaginable that the residents of Chankillo made a mistake when siting their fortress. Recent research has suggested that the key to understanding this place may lie not on the hilltop, but on the desert plain below.

Away from the ruined fortress and aligned north to south along the ridge of a nearby hill are thirteen towers. Recent excavations have uncovered further buildings to the east and west of the towers which archaeologists now believe to be intimately connected to this reptilian structure's true purpose. To see why, you must stand at the western observation point at the end of a night, facing the brightening eastern horizon through the towers. I have seen many sunrises, but nothing as dramatic and evocative as a Chankillo dawn. The edge of the solar disc, reddened and distorted by air heavy with sand, suddenly flares between two of the towers on the hill, and for the briefest of moments the Sun emerges as

a single sparkling diamond in the desert sky. Within seconds, the normally imperceptible rotation of our planet drags the star into full view, and you must avert your gaze as if to avoid staring into the face of a god.

The Thirteen Towers of Chankillo are more than a temple, however. It is thought that they are an ancient calendar, diligent timekeepers that have measured the passing of the days for thousands of years, outliving their creators by millennia. There is no clockwork here, no pendulums or cogs to keep the timepiece ticking; instead, time is measured using the most reliable pulse that the ancients had at their disposal – the Sun. In a beautiful piece of grand astronomical engineering, the thirteen towers are placed to mark the passing of time using the position of the sunrise on the eastern

The Thirteen Towers of Chankillo ... stand testament to our ancestor's instinct and desire to quantify and understand the ticking of the cosmic clock.

horizon. On 21 December, which in the Southern Hemisphere is the summer solstice – the longest day – the Sun rises just to the right of the most southerly tower, marking the beginning of a journey that will take it across the horizon as Earth orbits the Sun. As the year passes, the sunrise moves along the towers until, on 21 June – the shortest day – it rises just to the left of the northerly tower. So at any time of year, watching the sunrise at Chankillo would have allowed its inhabitants to determine the date within an accuracy of two or three days. I stood at the western observing point on 15 September, aware that the Sun has risen between the fifth and the sixth towers on this morning for the past two thousand years. Chankillo still works as a calendar because the Sun still rises and sets in very nearly the same places on the horizon today as it did when these stones were first set down.

Even though I understand the true nature of the Sun, when confronted with such a magnificent sunrise in such a dramatic and quiet place, I understand why these people would have almost certainly deified it. The high status of this place is clear, in that the scale of Chankillo is far grander than is necessary simply for a calendar. It is part-clock, part-temple, part-observatory; a place where on sacred days the people of Chankillo would have been able to greet the appearance of their god, the rising Sun, in the most spectacular of settings.

Today, the Thirteen Towers of Chankillo continue to tell the time, having long outlived their creators; they stand testament to our ancestors' instinct and desire to quantify and understand the ticking of the cosmic clock ◉

THE COSMIC CLOCK

Each day we awake to the rhythm of our planet as it spins at over 1,500 kilometres (932 miles) an hour, relentlessly rolling us in and out of the Sun's glare. Earth's ceaseless motion beats out the tempo of our lives with unerring repetition. A day is the twenty-four hours it takes Earth to rotate once on its axis; the 86,400 seconds it takes for anyone standing on the Equator to be whipped around the 40,074-kilometre (24,901-mile) circumference of our planet. This is the most obvious rhythm of the Earth, which comes about because of the spin rate of our rocky, iron-cored ball that was laid down somewhere in Earth's formation and 4.5-billion-year history.

Travelling at 108,000 kilometres (67,108 miles) an hour, we move through space in orbit around our star. Racing around the Sun at an average distance of 150 million kilometres (93 million miles), we complete one lap of our 970-million-kilometre (600-million-mile) journey in 365 days, five hours, 48 minutes and 46 seconds, returning regularly to an arbitrarily defined starting point. As we sweep through this place in space relative to the Sun, we mark the beginning and end of what we call a year.

Everywhere we look in the heavens we see celestial clocks marking the passage of time in rhythms. Our moon rotates around Earth every 27 days, seven hours and 43 minutes, and because it is tidally locked to Earth it also takes almost exactly the same amount of time to rotate on its own axis: 27 Earth days. This means that the Moon always presents the same face to Earth. Further out in the Solar System, a Martian day is very similar to our own, lasting one Earth day and an additional 37 minutes. But because Mars is further from the Sun, a Martian year lasts longer, with the red planet taking 687 Earth days

BELOW: Here on Earth our calendar is determined by the clockwork rhythm and movement of our planet as it rotates on its axis, working its way through space and along its annual orbit around the Sun.

to complete an orbit. In the farthest reaches of the Solar System, the length of a year gets progressively greater, with distant Neptune taking over 60,000 Earth days or 165 Earth years to make its way around its parent star. In September 2011, Neptune will have completed its first full orbit of the Sun since it was discovered in 1846.

As we look deep into space, the clockwork of the cosmos continues unabated, but as the distances extend, the cycles become grander, repeating on truly humbling timescales. Just as Earth and other planets mark out the passing of the years as they orbit the Sun, so our entire solar system traces out its own vast orbit. We are just one star system amongst at least 200 billion in our galaxy, and all these star systems are making their own individual journeys around the galactic centre. We are all in orbit around the super-massive black hole that lies at the heart of the Milky Way. It is estimated that it takes us about 225 million years, travelling at 792,000 kilometres (492, 125 miles) per hour to complete one circuit, a period of time known as a galactic year. Since Earth was formed four and a half billion years ago, our planet has made 20 trips around the galaxy, so Earth is 20 galactic years old. Since humans appeared on Earth a quarter of a million years ago, less than one-thousandth of a galactic year has slipped by. In Earth terms, that is the length of a summer's afternoon.

This is an immense amount of time; difficult to comprehend when we speak of the entire history of our species as the blink of a galactic eye. We live our lives in minutes, days, months and years, and to extend our feel for history across a galactic year is almost impossible. Yet here on Earth there are creatures that have existed for lengths of time that span these grandest of rhythms ◉

THE GALACTIC CLOCK

Nothing stays still in the Universe, our galactic clock is forever ticking, moving everything on to a new chapter in the story of the Universe, marking out the days, weeks, months and years in each and every planet in our galaxy. Everywhere in the heavens, time moves on using its own rhythms; as you journey to the farthest reaches of the Solar System the length of a year gets progressively greater and the cycles become grander. Every solar system among the 200 billion that exist in our galaxy makes its own unique journey around the galactic centre, as we all orbit the supermassive black hole that lies at the heart of the Milky Way Galaxy.

1 GALACTIC YEAR

MILKY WAY

SOLAR SYSTEM

Earth has completed 20 rotations / galactic years, since humans evolved we haven't yet completed 1 galactic day

365 DAYS

24 HOURS

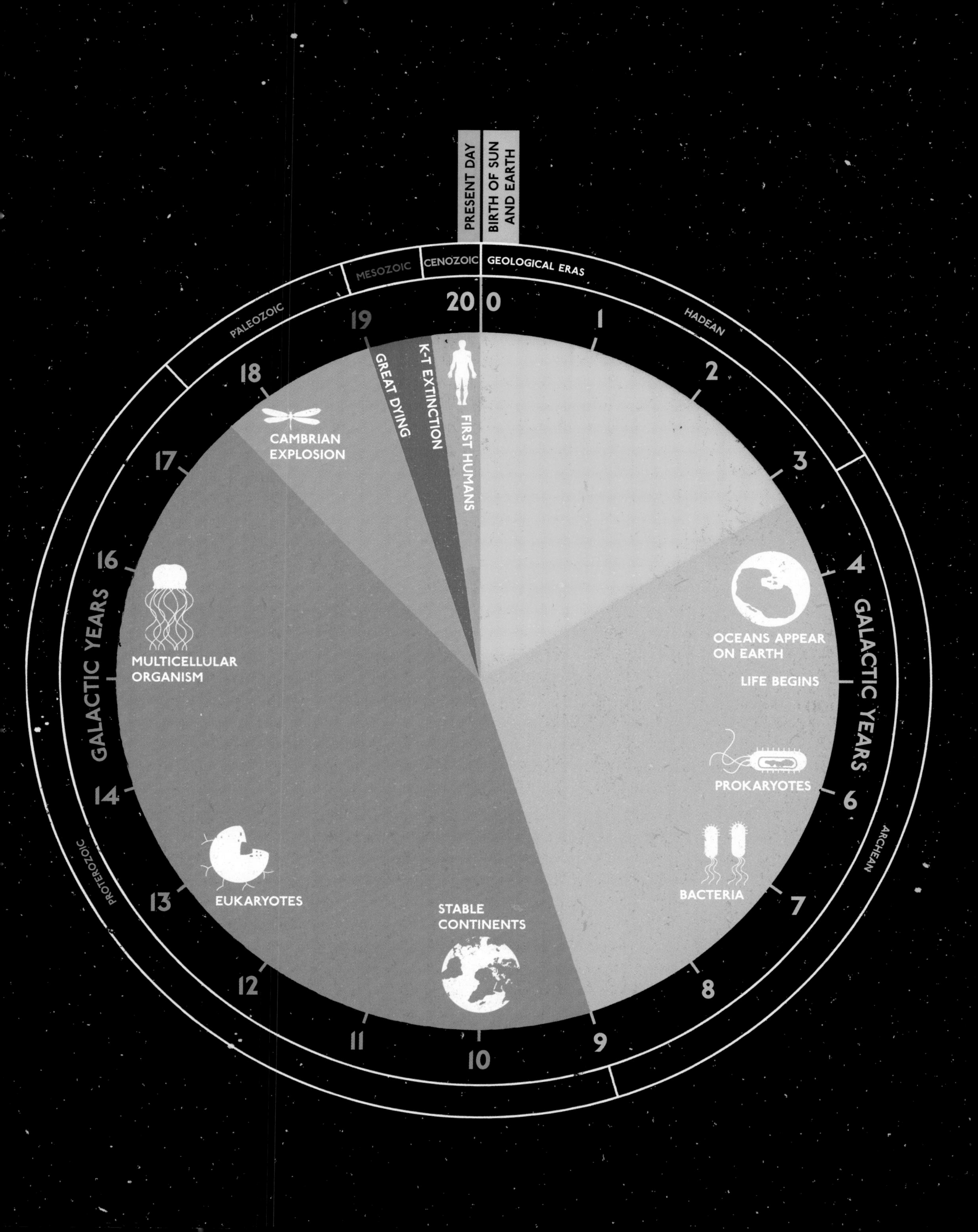

PRESENT DAY
BIRTH OF SUN AND EARTH
GEOLOGICAL ERAS
CENOZOIC
MESOZOIC
PALEOZOIC
PROTEROZOIC
ARCHEAN
HADEAN
GALACTIC YEARS
GALACTIC YEARS
20
0
1
2
3
4
6
7
8
9
10
11
12
13
14
16
17
18
19
FIRST HUMANS
K-T EXTINCTION
GREAT DYING
CAMBRIAN EXPLOSION
MULTICELLULAR ORGANISM
EUKARYOTES
STABLE CONTINENTS
BACTERIA
PROKARYOTES
LIFE BEGINS
OCEANS APPEAR ON EARTH

BELOW: Pregnant sea turtles return to the sands on the Pacific coast year after year in one of the oldest life cycles on Earth.

RIGHT: Galileo first investigated the physics of a swinging pendulum and how it could be used effectively for keeping time.

The Ostional wildlife refuge on the Pacific coast of Costa Rica is home to one of nature's most spectacular sights. On many nights of the year, a small number of tropical beaches along this thin land bridge between North and South America are visited by prehistoric creatures. They emerge from the ocean to lay their eggs in the sand. We filmed on Playa Ostional, a tiny strip of sand which is adjacent to a friendly village clustered around a makeshift football pitch. It is one of the few beaches in the world where large numbers of sea turtles make their nests, and the events that occur here form part of one of the oldest life cycles on Earth.

We are here to film the turtles hauling themselves from the ocean as they have done year on year without interruption for over 120 million years – half a galactic year. As we wait for them with our night-vision camera equipment, it is hard not to reflect on the sheer size of the mismatch in the histories of these ancient creatures and the species that built the football pitch by the sea. We humans know our planet well. We know there is a landmass called Europe, separated from Africa by a thin strip of ocean. We know that if you journey east from northern Europe you cross the vast expanses of Siberia and arrive eventually in Japan. Carry on, and you'll cross the Pacific Ocean and meet the Californian coast in the United States. The shape of our countries and continents is familiar and seemingly eternal, but the ancestors of the turtles I can see bobbing offshore were waiting for the right moment to crawl out onto the land when the shape of our continents was very different; they were waiting one hundred million years ago in the same ocean, but in those days the beaches marked out shorelines of continents that would be totally unrecognisable to our eyes. As the turtles patiently waited for their moment to give birth in the sand, the continents of Earth were slowly on the move. North America was close to Europe, South America was connected to Africa and Australia was joined with the Antarctic. It is moving to see the care with which these ancient creatures dig deep into the sand to protect their precious eggs, but equally powerful to reflect on the temporal mismatch between us and them. Collectively, they have witnessed the reshaping of our planet and the heavens above; the patterns of the stars must look very different from the other side of the Galaxy. I watch as one after another of these beautiful creatures covers its eggs and silently return to the ocean.

MEASURING TIME

Humans have long been measuring time, and we've developed our skills from the bluntest of temporal measurements to the extreme accuracy with which we can measure time today. The first attempts in chronometry may have begun thirty thousand years ago, when Stone Age humans used the lunar cycle to mark time. To early humans, the Moon would have marked out the clearest rhythm in the night sky, and by following it through its phases they were able to create the first calendars. Giving structure to the year beyond the day–night cycle allowed them to name periods of time, and so our classification and division of the cycles of the cosmos began.

Beyond the naming of the morning, afternoon and evening, the fine division of the day required the invention of one of our most enduring pieces of technology, the influence of which has been incalculable.

The first clocks were simple pieces of technology employed throughout the ancient world. Using nothing more complicated than a stick known as a 'gnomon' to cast a shadow, many civilisations were able to use sundials to track the passing of time during the day by measuring the movement of the shadow across a calibrated surface. Sundials are surprisingly accurate, but they have limited use as timekeepers, not least because they are difficult to use on a cloudy day and impossible to use at night!

Ancient Egypt was the first civilisation we know of that took measuring time beyond the sundial. The technique of using the flow of water to measure time may date as far back as 6000 BC, but the oldest physical evidence of a water clock can be found in the reign of Pharaoh Amenhotep III in 1400 BC. These elegant devices were simply stone vessels that allowed water to escape at a near-constant rate from a hole in the base. Inside the clock were twelve markings by which time could be measured as the water level dropped. These primitive clocks gave accurate measurements both night and day so that priests could perform their rituals at the appointed hour.

Water clocks continued to be refined and used by cultures across the globe for many centuries, and hourglasses employing the flow of sand to measure time were also used extensively. The Portuguese explorer Ferdinand Magellan used 18 hourglasses as a navigation tool on his ship when he circumnavigated the globe in 1522.

Time keeping was elevated to a completely new level of accuracy with the invention of pendulum clocks. Galileo

Christiaan Huygens invented the first pendulum clock in 1656, and it remained the most accurate way of telling the time until the 1930s.

was the first scientist to investigate the physics of a swinging pendulum. The key property of the pendulum, which makes it useful as a timekeeping device, is that the period of the swing – the familiar tick-tock of the clock – depends only on the length of the pendulum and Earth's gravitational pull. Perhaps counterintuitively, the period doesn't depend on how high you lift the pendulum to start the swing, as long as it's not too high. Physics students have the formula for the time period of a pendulum permanently etched in their minds. It is:

$$T \approx 2\pi\sqrt{\frac{L}{g}}$$

where T is the period, L is the length and g is the acceleration due to gravity – in other words, a measure of the strength of Earth's gravitational field, which is almost the same wherever you are on Earth; approximately 9.81 metres (300

feet) per second squared. This means that all you need to do to make a clock that ticks accurately is get the length of the pendulum right. Most grandfather clocks have a pendulum that swings with a period of two seconds, which a little simple mathematics will tell you requires a pendulum approximately one metre long. The Dutch astronomer Christiaan Huygens invented the first pendulum clock in 1656, and it remained the most accurate way of telling the time until the 1930s.

Today we rely on atomic clocks to measure time with extraordinary accuracy. Atomic clocks use the frequency of light emitted when electrons jump around in atoms (usually caesium) as the 'pendulum'. This is highly accurate because the structure of atoms is unchanging, and therefore the light emitted from them always has the same frequency. This light can be used, with some clever engineering, to keep an oscillator ticking at a precise rate, allowing atomic clocks to tell the time with an accuracy of one-thousand-millionth of a second per day. The second itself has been defined since 1967 using the theory behind atomic clocks; one second is defined as the duration of 9,192,631,770 periods of the radiation corresponding to the transition between the two hyperfine levels of the ground state of the caesium 133 atom. In English, this means a second is the time it takes for 9,192,631,770 peaks in a wave of light, emitted when an electron makes a specific jump in an atom of caesium, to fly past you.

Atomic clocks allow us to measure incredibly small periods of time. Until now, the shortest period we have been able to measure is 12 attoseconds, or 12 quadrillionths of a second. This is how long it takes light to travel past 36 hydrogen atoms lined up together. That's not far at all ◉

For all the accuracy and precision we have achieved in keeping time, we have never managed to do anything more than observe it. From the very earliest solar calendars to the electrons jumping around in caesium atoms, one thing about the nature of time is clear: we can measure its passing, but we cannot control it. It moves inexorably forward; it cannot be stopped. This tells us something profound about our universe.

RIGHT: The Perito Moreno glacier in Patagonia, Southern Argentina, is a stark but beautiful place where the passage of time moves progressively forward but so slowly that it almost goes unnoticed.

THE ARROW OF TIME

Few places on our planet are as spectacular as the Perito Moreno glacier in Patagonia, southern Argentina. This dense blue wall of frozen water in the Los Glaciares National Park is part of a system of hundreds of glaciers that sweep down the continent from the southern Patagonian ice fields. Together they form the third-largest icecap on our planet. The Perito Moreno glacier alone covers an area of 250 square kilometres (96 square miles) and in places it is 170 metres (560 feet) deep. The ice ends where solid meets liquid at Lake Argentino; a great wall of ice towers over the surface of the lake, and the few who make it to this bleak but utterly beautiful place have the chance to sail along its edge across one of the most dramatic expanses of water in the world.

At first sight the glacier appears static and unmoving; standing on the lake shore, this seems like a place where the passage of time goes as unnoticed as the laws of physics will allow. Yet there is a reason why boats don't venture too close to the edge of the ice cliff. As we approached I didn't only see the passage of time; I felt it. This glacier is in constant motion; relentlessly carving its way down from the Andes as it has done for tens of thousands of years. At the glacier's edge, the wall of ice is 70 metres (230 feet) high, and the whole face of the glacier is sliding into the lake at around 50 centimetres (20 inches) per day. That means that well over a quarter of a billion tonnes of ice cascades into the lake every year. You don't often see it, but you can hear it; every now and then there is a tremendous cracking sound, followed by a deep rumbling. The surface of the lake comes alive as a turbulent wave powers beneath your boat. The pace of change in this place is anything but glacial. It is so vast and complex that you perceive it to be alive; an unpredictable, overwhelmingly powerful organism clawing the land in vain as it inevitably slides into the waters.

This is all part of a highly ordered sequence. As time passes, snow falls, ice forms, the glacier gradually inches down the valley, and when the ice meets the water, pieces break off and fall into the lake creating waves. In many ways

We expect to see ice fall from the glacier, splash into the water and create waves. If it happened in any other way we'd immediately know there was something wrong.

LEFT AND BELOW: A great wall of ice towers over the surface of Lake Argentino, where this vast, seemingly immovable glacier is slowly and relentlessly sliding down into the icy waters below.

this ordering of events into a sequence is the simplest way to think about time. The fact that sequences of events always happen in order is a fundamental part of our experience of the world. We expect to see ice fall from the glacier, splash into the water and create waves. If it happened in any other way we'd immediately know there was something wrong. Yet there is a legitimate question here about what we mean by events happening 'in order'. However long we might stand on the edge of this beautiful lake we would never expect to see this dramatic sequence of events happen in reverse, even though there is nothing in the laws of nature that prevents this happening. There is no physical reason why all the water molecules moving around in the lake shouldn't gather together on the surface, reduce their collective temperature such that they bind together to form ice, jump out of the water and glue themselves onto the surface of the glacier. We do, however, have a scientific explanation for why such a dramatic reversal never happens; we call it the 'arrow of time'.

This phrase was first used by the British physicist Sir Arthur Eddington in the early twentieth century to describe this deceptively simple and yet profound quality of our universe: it always seems to run in a particular direction. Eddington was instrumental in bringing Einstein's theory of relativity to the English-speaking world during the First World War, and also one of the first scientists to directly confirm the findings of relativity when he led an expedition to observe the total solar eclipse on 29 May 1919. In 1928 he published *The Nature of the Physical World*, in which he introduced two great ideas that have endured in popular scientific culture to this day. The first was the image of the infinite monkey theorem, which states that given an infinite amount of time, anything consistent with the laws of physics will happen: 'If an army of monkeys were strumming on typewriters, they might write all the books in the British Museum'. This is related in a deep way to the arrow of time, which Eddington described as follows:

'Let us draw an arrow arbitrarily. If as we follow the arrow we find more and more of the random element in the state of the world, then the arrow is pointing towards the future; if the random element decreases the arrow points towards the past. That is the only distinction known to physics. This follows at once if our fundamental contention is admitted that the introduction of randomness is the only thing that cannot be undone. I shall use the phase "time's arrow" to express this one-way property of time which has no analogue in space.'

Eddington's arrow vividly and economically expresses a key property of time; it only goes in one direction. But what does he mean by randomness? It seems obvious that the Universe is constantly evolving, but what drives this evolution? How should we quantify how random something is? Why is the past different from the future? Why is there an arrow of time? Time is something we all understand, and yet a plausible scientific reason as to why time marches inexorably forward wasn't offered until the late nineteenth century, coming about as the solution to a practical problem on Earth ◉

THE ORDER OF DISORDER

BELOW: In a series of simple experiments, Joule demonstrated that mechanical work could be converted into heat. Using a paddle wheel turned by falling weights, he stirred water in an insulated barrel and observed how the temperature of the water rose by the amount that depended on how far the weights fell.

In 1712 the English inventor Sir Thomas Newcomen created the first commercially successful steam engine, paving the way for the Industrial Revolution. This accolade is more usually awarded to the Scottish inventor James Watt. In 1763 Watt was asked to repair a Newcomen engine by the University of Glasgow, and in doing so he developed a new steam engine which, it is appropriate to say without hyperbole, transformed the landscape of modern life. Watt's steam engine was more efficient and more flexible than its predecessor; it used far less coal than the Newcomen for a given power output, and was therefore much cheaper to run. More importantly still, Watt's engine could do more than pump water out of the wet mines, it could also generate the rotary motion that was needed to power the machines on the factory floor. No longer did a factory have to be situated by a river to turn its equipment; with the help of Watt's engine a factory could be sited anywhere, catalysing the emergence of the modern industrial landscape. Steam-powered machines changed the course of history, and yet despite their importance, the nineteenth-century engineers who followed Watt struggled to improve them. There seemed to be fundamental principles that restricted their efficiency, but with profit margins to maximise, even a small increase in their effectiveness would be highly valuable. So understanding how hot the fire should be or what substance should be boiled in the engine were problems that were not only interesting from a scientific perspective but were also critical for businesses. It was out of these questions of engineering design that the science of thermodynamics arose, and with it the concepts of heat, temperature and energy entered the scientific vocabulary in a precise way for the first time.

One of the scientists working on these problems was the German mathematician Rudolf Clausius. Clausius was interested in heat, which until the first half of the nineteenth century was thought to be a fluid that flowed from hot things to cold things. Clausius and others realised that this description was not able to explain the cycle of a steam engine. The foundation for Clausius's theoretical advances was laid by one of his contemporaries, the English physicist and brewer James Joule, who was working to improve the efficiency of the steam engines in his brewery. What finer motivation for the advance of fundamental physics? The quest for cheaper beer motivated him to investigate the relationship between the work his steam engines could do, and heat. In doing so he managed to reduce the costs of beer production and lay one of the cornerstones of the science of thermodynamics.

Using a series of beautifully simple experiments, Joule was able to demonstrate that mechanical work could be converted into heat. One such experiment used a falling weight to spin a paddle within an insulated barrel of water. Joule knew how much work was done by the falling weight and so could measure the temperature rise of the water. He conducted similar experiments on compressed gases and flowing water, and each time he found that it took the same amount of work to raise the temperature of a fixed amount of water by one degree Fahrenheit. Inscribed on his tombstone in Brooklands cemetery near Manchester is the number 772.55 – his measurement of the amount of work done in foot-pounds force that is required to raise the temperature of one pound of water by one degree Fahrenheit.

The reason that Joule's work was important is that it demonstrated that heat is not a thing that can be created or destroyed. It doesn't literally flow between things or move around, it is in fact a measure of something else. Even today, this is perhaps not obvious because we still speak of the flow of heat from hot to cold things. Heat, we now understand, is simply a form of energy. Just as a ball resting on a table has energy which can be released by dropping it (known as gravitational potential energy), so a

$$C\Delta T = Mgh$$

- M is the mass of the falling weight
- g is the acceleration due to gravity
- h is the distance through which the weight falls
- C is 'the specific heat capacity of water' (the amount of heat required to heat one kilogramme of water by 1°)
- ΔT is the actual rise in temperature caused by the stirring

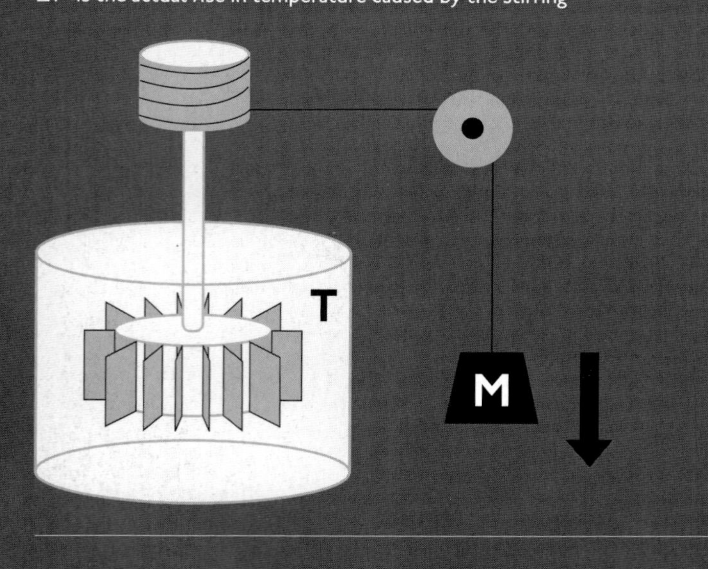

BELOW: Newcomen's engine, created in 1712, was the first commercially successful steam engine and laid the foundations for the work of other inventors, such as James Watt, which would power forward the Industrial Revolution in Britain. The Newcomen atmospheric engine was used to pump water out of coal mines, using a pivoted arm (top) to transfer power between the piston and the rod. The piston was driven down by the pressure of a partial vacuum in the cylinder, which drew the rod upwards. As steam in the cylinder condensed, the piston was forced up, and the rod down.

hot thing has energy that can be released, at least in part, by putting it next to a cold thing. To heat something up, you simply have to transfer energy to it by doing work on it, as Joule found by using a falling weight, and it doesn't matter how that work is done. It can be a falling weight, a shining light or an electric current, but as long as you do the same amount of work, the temperature increase will be the same. This was all quantified, as a result of Joule's work, into the First Law of Thermodynamics, which is a statement of the fact that energy cannot be created or destroyed; it can only be changed from one form into another. Rudolf Clausius made the first explicit statement of the law, and laid down the foundations of the science of thermodynamics, in his landmark 1850 publication 'On the mechanical theory of heat'.

The first law can be written down mathematically as

$$\Delta U = Q - W$$

which in words says that the increase in the internal energy of something (ΔU) is equal to the heat flow into it (Q) minus the work performed by it (W). If you performed work on it, the W would have a plus sign, and if you took heat out of it, the Q would have a minus sign.

Fifteen years after writing down the first law of thermodynamics, and far more importantly for our understanding of the arrow of time, Clausius introduced a new concept known as entropy, which lies at the heart of the Second Law of Thermodynamics. Clausius's statement of the second law does not at first sight sound as if it has profound implications for the future of our universe. He simply stated that 'No process is possible whose sole result is the transfer of heat from a body of lower temperature to a body of higher temperature'. This simple proposition occupies such a profound position in modern science that Arthur Eddington said of the second law:

'If someone points out to you that your pet theory of the Universe is in disagreement with Maxwell's equations, then so much the worse for Maxwell's equations. If it is found to be contradicted by observation, well, these experimentalists do bungle things sometimes. But if your theory is found to be against the Second Law of Thermodynamics I can give you no hope; there is nothing for it but to collapse in deepest humiliation.'

The concept of entropy enters when the second law is written down in quantitative form. The change in entropy of a system, such as a tank of water, is simply the amount of heat added to it at a fixed temperature. In symbols,

$$\Delta S = \frac{\Delta Q}{T}$$

where ΔS is the change in the entropy as a result of adding a small amount of heat, ΔQ, at a fixed temperature T. It may still be unclear what this has to do with the Universe, but here is the profound point discovered by Clausius. In any physical process at all, you find that entropy either stays the same or increases. It *never* decreases. Here is the thermodynamic arrow of time. Clausius had discovered a physical quantity that can be measured and quantified which only ever increases in practice, and never decreases even in theory, no matter how cleverly you design your experiment or piece of machinery. This is extremely useful information if you are designing a steam engine, because it puts a fundamental limit on the efficiency. It also prevents the construction of the so-called 'perpetual motion machines' so beloved of crackpot inventors to this day. You could say that the second law tells you that you can't get something for nothing, but the second law is more profound than this, because it introduces a difference between the past and the future. In the future, entropy will be higher than it is in the present because it always increases. In the past, entropy was lower than it is now because it always increases.

Clausius introduced the concept of entropy because he found it useful, but what exactly is entropy, and what is the deep reason that it always increases? And what was the meaning of Eddington's cryptic quote about randomness and the arrow of time? He seemed to be equating entropy with the amount of randomness in the world, and indeed he was. Understanding this will make it clear why the Second Law of Thermodynamics mandates that our entire universe must, one day, die ◉

ENTROPY IN ACTION

In 1908 in the small town of Kolmanskop in southern Namibia, a railway worker by the name of Zacharias Lewala found a single diamond lying in the sand. He showed the precious stone to his manager – railway inspector August Stauch – who immediately realised its significance and set in motion a train of events that turned this desolate place into one of the most valuable diamond mines in the world. The colonial German government closed the entire area to outsiders; only German entrepreneurs were allowed to make their fortunes here. For 40 years, Kolmanskop was home to a thriving community as over a thousand people gathered, seeking to become millionaires by picking diamonds out of the desert. As the money rolled in, the residents built a town in the finest German tradition; grand houses stood beside a casino, a ballroom and the first X-ray station in the Southern Hemisphere. They led a champagne lifestyle in the desert, and created a little piece of opulent German architecture in the sand. Eventually, though, as with all cash cows, the diamonds could no longer be found and the town gradually lost its sparkle until it was abandoned in 1954. For half a century it has fallen into disrepair as the buildings are slowly reclaimed by the sands.

Today Kolmanskop is a ghost town, a place where our efforts to replace the geological grandeur of the desert with architectural grandeur of our own have been thwarted by the power of the winds.

Kolmanskop lies just outside the modern port town of Lüderitz, which sits in spectacular isolation on the southern Namibian coast. One of our guides told us that it takes a special kind of Namibian to set up home in Lüderitz – you have to really want to live there. The reason this place has a reputation for being particularly harsh, even by the standards of this part of the world, is the wind. This strip of the southern African

BELOW: The opulent buildings of the once-glorious town of Kolmanskop are a shadow of their former selves as the desert sands blow across them and reclaim the landscape.

coast is permanently assaulted by the untamed winds of the South Atlantic that whip up the fine-grained sands of the Namib Desert and hurl them unrelentingly into machinery, houses, camera equipment and eyes. I have never experienced anything like it. While filming, I found myself walking through the wind at Kolmanskop with my hands completely shielding my face. I didn't do this for dramatic effect, I genuinely couldn't look into the lacerating sand-laden wind. We also shot a scene showing a little sandcastle gradually blowing away; the camera we left in the desert for hours to film that had its lens sandblasted – the high-precision optics felt like sandpaper after a single spring afternoon in the vicinity of Lüderitz. If it wasn't for the fact that it never rains here, and nothing rots, the ghost town of Kolmanskop would surely already have disappeared back into the desert.

The little sandcastle slowly decaying in the desert wind vividly demonstrates the connection between decay, randomness and entropy. To understand why this is so, we'll need a different and much more intuitive definition of entropy than that given by Clausius. Known as the statistical definition of entropy, it was developed by Ludwig Boltzmann in the 1870s.

A sandcastle is made of lots of little grains of sand, arranged into a distinctive shape – a castle. Let's say there are a million sand grains in our little castle. We could take those million grains and, instead of carefully ordering them into a castle, we could just drop them onto the ground. They would then form a pile of sand. We would be surprised, to say the least, if we dropped our sand grains onto the floor and they assembled themselves into a castle, but why does this not happen? What is the difference between a pile of sand and a sandcastle? They both have the same number of sand grains, and both shapes are obviously possible arrangements of the grains. Boltzmann's definition of entropy is essentially a mathematical description of the difference between a sandcastle and a sand pile. It says that the entropy of something is the number of ways in which you can rearrange its constituent parts and not notice that you've done so. For a sandcastle, the number of ways in which you can arrange the grains and still keep the highly-ordered shape of the castle is quite low, so it therefore has low entropy. For a sand pile, on the other hand, pretty much anything you do to it will still result in there being a pile of sand in the desert, indistinguishable from any other pile of sand. The sand pile therefore has a higher entropy than the sandcastle, simply because there are many more ways of arranging the grains of sand such that they form a pile of sand than arranging them into a castle. Boltzmann wrote this down in a simple equation, which is written on his gravestone:

$$S = k_B \ln W$$

S is the entropy, W is the number of ways in which you can arrange the component bits of something such that it is not changed, and k_B is a number known as Boltzmann's constant. For the more mathematically adventurous, ln stands for 'natural logarithm'. If you don't know what that means, don't worry; the equation simply relates to the number of ways in which you can arrange things to the entropy.

As long as each particular arrangement of the sand grains is equally likely, then if you start moving sand grains around at random they are overwhelmingly more likely to form a shapeless pile of sand than a sandcastle.

That may seem a bit complicated, and not entirely illuminating yet, but here is the key point: as long as each particular arrangement of the sand grains is equally likely, then if you start moving sand grains around at random they are overwhelmingly more likely to form a shapeless pile of sand than a sandcastle. This is because most of the arrangements you create at random look like a formless pile, and very few look like a sandcastle.

This is common sense, of course, but now think about what this looks like at a microscopic level – the level of individual sand grains. There is nothing at all in the laws of nature to stop the wind blowing a grain of sand off one of the turrets of our castle and then picking up another grain from the desert and blowing it back onto the turret again, leaving our castle perfectly unchanged. Nothing at all, that is, other than pure chance. It is much more likely that the grains of sand blown off the castle are not replaced with others from the desert, and so our castle gradually disintegrates, which is to say it gradually changes into a formless sand pile. In Boltzmann's language, this is simply the statement that the entropy of the castle will increase over time; the castle will become more and more like a sand pile. Why? Because there are many more ways of arranging the grains of sand into a pile than there are into a castle, so if you just randomly blow grains around they will tend to form piles more often than castles. Here is the deep reason that entropy always increases: it's simply more likely that it will! Notice that there is nothing in the laws of nature that prevent it from decreasing; it's possible that the wind will build a sandcastle, but the chances are akin to tossing a coin billions of times and each one coming up heads. It's simply not going to happen.

LEFT AND BELOW: Watching my carefully constructed sandcastle gradually disintegrate in the strong desert winds perfectly demonstrates how entropy increases, and the idea that the past is always more ordered than the future.

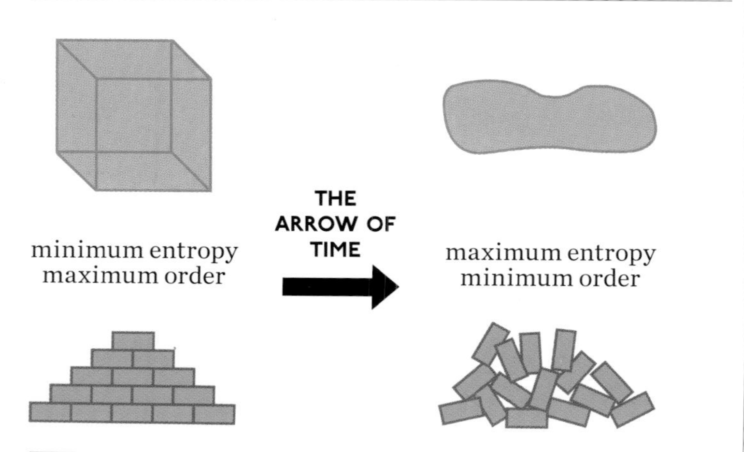

Boltzmann's statistical definition of entropy is the key to understanding Eddington's arrow of time. This is such a key concept with such profound consequences that it is worth repeating it once more in a slightly different way. If there are a million different ways of arranging a handful of sand grains, with 999,999 of the ways producing disordered sand piles but only 1 producing a beautifully ordered castle, then if you keep throwing the sand grains up in the air they will usually land in the form of a disordered pile. So, over time, if there is a force like the wind that acts to rearrange things, things will get more messy or disordered simply because there are more ways of being disordered than ordered. This means that there is a difference between the past and the future: the past was more ordered and the future will be less ordered, because this is the most likely way for things to play out. This is what Eddington meant by his statement that the future is more random than the past, and his description of the arrow of time as the thing that points in the direction of increasing randomness. And this is why entropy always increases.

For the purposes of our story, this is sufficient; if you take a university physics degree, this is what you will learn about entropy and the arrow of time. But there is still a great deal of debate and research surrounding entropy, and it centres on something we have dodged slightly. We have only spoken about entropy differences; the past had a lower entropy than the future; ordered things become disordered as time ticks by, but one might legitimately ask where all the order in the Universe came from in the first place. In the case of our sandcastle, it's obvious – I made it – but how did I get here? I'm very ordered. How did Earth get here? It's very ordered too. And how did the Milky Way appear if it is composed of billions of ordered worlds orbiting around billions of ordered stars? There must have been some reason why the Universe began in such a highly ordered state, such that it can gradually fall to bits. The answer is that we don't know why the Universe began with sufficient order in the bank to allow planets, stars and galaxies to appear. We understand how gravity can create local order in the form of solar systems and stars, but this must be at the expense of creating more disorder somewhere else. So there must have been a lot of order to begin with. In other words, the Universe was born in a highly ordered state, and there should be a reason for that. It is unlikely to have been chance, because by definition a highly ordered state is less likely to pop into existence than a less ordered one; a sandcastle is less likely to be formed by the desert winds than a pile of sand. Since the Universe is far less ordered today than it was 13.75 billion years ago, this means it is far more likely that our universe popped into existence a billionth of a second ago, fully formed with planets, stars, galaxies and people, than it is that the Universe popped into existence at the Big Bang in a highly ordered state. There is clearly something fascinating about the entropy of the early Universe that we have yet to understand ◉

The arrow of time has been playing out dramatically in Kolmanskop since the mining facility was abandoned in 1954. In every building you can see the gradual transition from order to disorder; every room that was once full of structure is slowly being returned to a less-ordered state. This is the march of the arrow of time on Earth, but it is nothing compared to the grand journey that time's arrow forces our universe to make.

LEFT: The sands of time are slowly and literally overrunning Kolmanskop, dismantling the highly crafted town and returning it to dust once more.

THE LIFE CYCLE OF THE UNIVERSE

Our Universe follows the law of any living thing: it develops in stages from birth through life and ultimately to death. We understand the early stages of its life because observations by scientists have provided valuable information as to how the Universe was created, and also fill in the crucial facts about the history of the Universe thus far. We are living in an early phase of our universe, the Stelliferous Era, with many more stages of life and change still to come, and yet we can confidently make predictions about our Universe's future. By observing the life cycles of the stars above us we can map out the remaining years of our universe's life.

1. PRIMORDIAL

$0 - 10^5$

Big Bang, inflation, and nucleosynthesis take place. Towards end of this era the Universe becomes transparent for the first time

2. STELLIFEROUS

$10^6 - 10^{14}$

Our current era. Matter is arranged in stars, galaxies, and galaxy clusters

3. DEGENERATE

$10^{15} - 10^{35}$

Galaxies no longer exist. This is the era of brown dwarfs, white dwarfs, and black holes. The Sun becomes a black dwarf. White dwarfs will assimilate dark matter and continue with minimal energy output

4. BLACK HOLE ERA

$10^{40} - 10^{100}$

Organized matter will remain only in the form of black holes. Black holes slowly 'evaporate' away the matter contained in them. By the end of this era, only extremely low-energy photons, electrons, positrons, and neutrinos will remain

Unstable Positronium Atom

Positron

ULTIMATE FATE OF THE UNIVERSE: THE THEORIES

5. DARK ERA

$10^{101} - \infty$

The Universe will be nearly empty. Photons, neutrinos, electrons, and positrons will fly from place to place. Electrons and positrons will occasionally form positronium atoms. These structures are unstable, however, and their constituent element will eventually annihilate each other

Neutrino

6. HEAT DEATH

Considered to be the most likely fate of the Universe. It will occur if the Universe continues expanding as it has been. The continued expansion will result in a universe that approaches absolute zero temperature

6. BIG BOUNCE

This is a cyclical repetition interpretation of the Big Bang whereby the first cosmological event was the result of the collapse of a previous universe

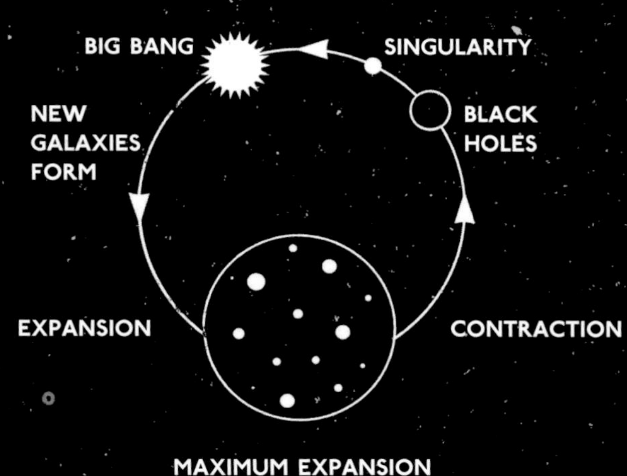

6. BIG CRUNCH

The expansion of space will reverse and the Universe will re-collapse, ultimately ending as a black hole singularity

EXPANSION

BIG BANG SINGULARITY

BIG CRUNCH SINGULARITY

TIME

6. MULTIVERSE

Our universe is merely one Big Bang among an infinite number of simultaneously expanding Big Bangs that are spread out over endless distances

THE LIFE OF THE UNIVERSE

BELOW: The Sun is one of at least two hundred billion stars in our galaxy, and it, along with countless others, shine brightly over Earth, night and day, in an ever-changing, ever-evolving cosmos.

RIGHT: In our age of stars, the Milky Way Galaxy is filled with stars igniting and scattering their light across the night sky.

Just as human beings, planets and stars are born, live their lives and die, so the Universe also lives its life in distinct stages. It began 13.75 billion years ago with the Big Bang, and in this embryonic period, known as the Primordial Era, the Universe was a place without the light from the stars, although in its early years the swirling hot matter would have glowed as brightly as a sun. For the first 100 million years, the conditions were far too violent for stars to form. This changed when the Universe had expanded and cooled sufficiently for the weak force of gravity to begin to clump the primordial dust, gas and dark matter into galaxies. With this came the dawning of the second great epoch in the life of our universe: the Stelliferous Era, the age of stars.

The moment the first stars were born is one of the most evocative milestones in the evolution of the cosmos. It signals the end of an alien time when the Universe was without structure – a formless void. The beginning of the Stelliferous Era marks the beginning of the age of light, the moment when the Universe would have become recognisable to us. The sky would have become black, punctuated with the glowing mist of the galaxies and the sharp silver of the stars. This is our universe today – a place where starlight decorates our nights and illuminates our days.

Our sun is one of at least two hundred billion stars in our galaxy; one of a hundred billion galaxies in the observable universe. We live in a cosmos of countless islands of countless stars which bathe the Universe in light. Yet despite the fact that the Universe is over 13 billion years old, we are still just at the beginning. Although the cosmos is awash with stars, is populated with vast nebulae and systems of planets and countless billions of worlds that we've yet to explore, we are living close to the beginning of the Stelliferous Era, an era of astonishing beauty and complexity. But the cosmos isn't static and unchanging; it won't always be this way because as the arrow of time plays out, it produces a cosmos that is as dynamic as it is beautiful.

The moment the first stars were born is one of the most evocative milestones in the evolution of the cosmos ... it marks the beginning of the age of light, the moment when the Universe would have become recognisable to us.

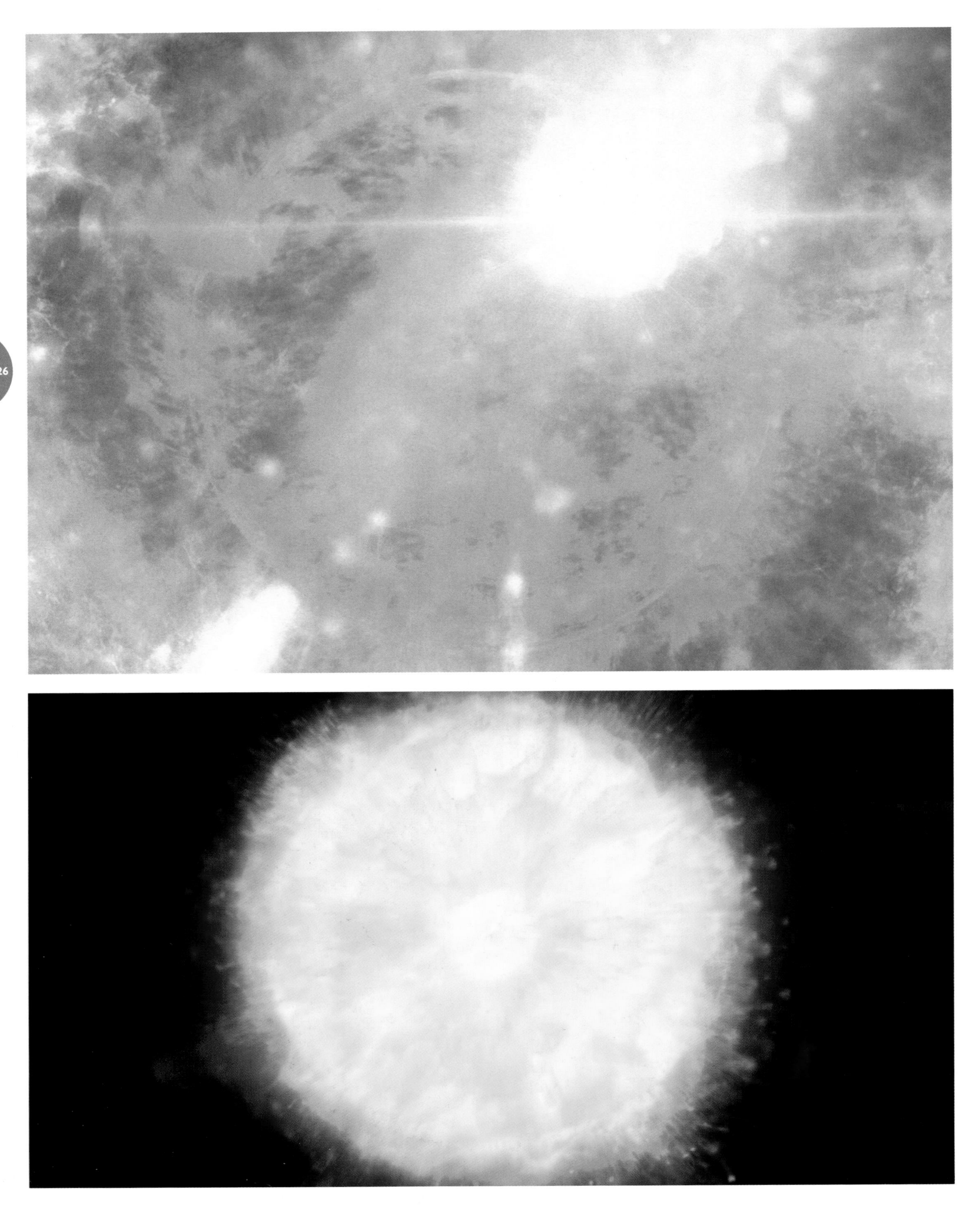

LEFT: A gamma-ray burst is one of the Universe's most spectacular and luminous explosions. As the core of a dying star collapses into a black hole, gas jets blast out from it into space.

BELOW: This dramatic image shows the gamma-ray burst from GRB 090423, combining data from the Ultraviolet/Optical (blue, green) and X-ray (orange, red) telescopes of NASA's Swift satellite.

When these stars run out of nuclear fuel … they die in a dramatic fashion, collapsing in an instant and releasing more energy in one second than our sun will produce in its entire 10-billion-year lifetime.

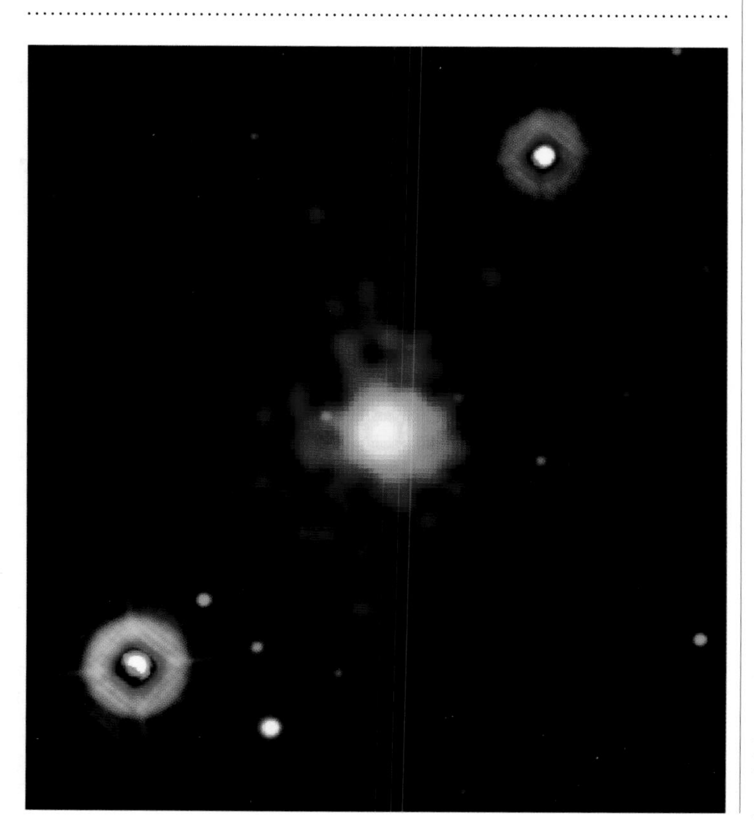

THE FIRST STAR

On 23 April 2009 at 07.55 GMT, NASA's Swift detected one of the most distant cosmic explosions ever seen – a gamma-ray burst that lasted ten seconds. The Swift satellite was designed and built with the intention that it would aid the study of a rare type of event known as a gamma-ray burst. These events, which last only a few seconds, are the most energetic and powerful emitters of radiation in the known universe. It is thought that gamma-ray bursts occur in supernova explosions – as the dying act of the most massive stars as they collapse to form black holes. By 08.16 GMT, minutes after the burst had faded away, the UK's Infrared Telescope (UKIRT) in Hawaii saw the glowing ember of the explosion. As the day wore on, the largest telescopes across the world focused on the event as it appeared above their horizon. The afterglow was observed for several hours, but by 28 April the event had faded completely from view.

The picture shown here merges data from two of Swift's telescopes, and the important feature of this composite image is the rather unremarkable-looking red blob at the centre. This blob is the fading remains of GRB 090423 – once one of the brightest stars in the Universe. The poetically named GRB 090423 was once a Wolf-Rayet star. Named after the two French astronomers who discovered the first one in 1867, Wolf-Rayet stars are massive – over twenty times the mass of our sun – and because they are so massive, and burn so brightly, they are also extremely short-lived. When these stars run out of nuclear fuel after only a few hundred thousand years, they die in a dramatic fashion, collapsing in an instant and releasing more energy in one second than our sun will produce in its entire 10-billion-year lifetime.

GRB 090423 was a big Wolf-Rayet star – perhaps 40 or 50 times the mass of the Sun – however, this is not the only thing that is interesting about it. It's not just the story of the death of this star, revealed by the brief appearance of the pale red dot, that has captivated astronomers, it's the age of it. The light from this dot has travelled a very long way across the Universe to reach us, and has taken a very long time to do it. When we look at the afterglow of this explosion, we are looking at an event that happened a long time ago, in a galaxy far, far away. In fact, this light has been travelling towards us for almost the entire history of the Universe. GRB 090423 died over thirteen billion years ago, just over 600 million years after the Universe began. This is incredibly early in the Universe's history. At the time of filming *Wonders of the Universe*, in autumn 2010, GRB 090423 was the oldest single object ever seen, although just after filming a galaxy was discovered in the Hubble Space Telescope's Ultra Deep Field Image (see pages 54–55) that is slightly older than GRB 090423. Even more poetically named UDFy-38135539, this galaxy currently holds the distance and age record with a light travel time of slightly over 13 billion years. Allowing for the expansion of the Universe, the (so-called co-moving) distance of UDFy-38135539 is currently 30 billion light years away from Earth.

However, it is the discovery of GRB 090423, this ghostly pale red dot, and the sight of the explosive death of one of the first stars in the Universe, that gives us a glimpse of the grandest timescale of them all ◉

THE DESTINY OF STARS

The arrow of time has been playing out in every corner of the Universe since the beginning of time. It dictates the destiny of everything; our civilisation, our planet, the Solar System, and all that lies beyond. The entropic march is inevitable and relentless. Nothing can resist the arrow of time, nothing can last forever, no star can shine without end and no planet can continue to turn. The Universe, bound by the laws of nature, must decay towards a radically different tomorrow.

BELOW: We take for granted the sight of the Sun rising and setting on our horizon, but we now know its presence is not eternal.

BOTTOM: The computer-generated image shows how dramatically different the Sun will look in our heavens as it dies and dims.

Today, 13.7 billion years after the Universe began, we are living through the most productive era that our universe will ever know. The Stelliferous Era is a time of life and death, with the constant dance between gravity and nuclear fusion creating a dynamic, ever-changing landscape in the heavens. For a human being, for whom a century is a lifetime, the changes may appear slow, but be in no doubt that you are part of the Universe at its most vibrant. As we've watched the stories of stars like GRB 090423 play themselves out in the night sky, we have seen at first hand that no star can last forever. Every one of those brightly burning lights has a destiny as defined and as certain as our own, and this of course includes the star at the centre of our solar system.

The Sun was formed 4.57 billion years ago from a collapsing cloud of hydrogen and helium and a sprinkling of heavier elements. For the tiniest fraction of this time, humans have marked the passing of the days as it rose and set, and surely considered it to be an eternal presence. It was only during the twentieth century that we discovered the Sun's fires must one day dim ◉

THE DEMISE OF OUR UNIVERSE

RIGHT, TOP: An artist's impression of Sirius A and its diminutive companion Sirius B in close-up. They are overlain on a real image of the night sky containing the three stars of the Summer Triangle: Vega, Deneb and Altair. As seen from Sirius, our sun would appear as a moderately bright star in this same area of sky. It is shown here just below right of Sirius A.

RIGHT, BOTTOM LEFT: This image from NASA's Hubble Space Telescope shows the Boomerang Nebula in early 2005 and the two lobes of matter that are being ejected from the star as it dies. The rapid expansion of the planetary nebula around this dying star has made it one of the coldest places found in the Universe so far.

RIGHT, BOTTOM RIGHT: A Hubble Space Telescope image of the dazzling Sirius A with the faint speck of Sirius B to its lower left. Sirius B is 10,000 times fainter than Sirius itself.

At the moment the Sun is in the middle of its life, fusing hydrogen into helium at a rate of around 600 million tonnes every second. It will continue to do this for another five billion years; but eventually, perhaps fittingly given the grandeur and beauty it has nurtured in its empire, it won't simply fade away. As the stores of hydrogen run dry, the Sun's core will collapse and momentarily, as helium begins to fuse into oxygen and carbon, a last release of energy will cause its outer layers to expand. Imperceptibly at first, the extra heat of the Sun will extend towards us as its diameter increases by around 250 times. The fiery surface of our star will move beyond Mercury, towards Venus and onwards to our fragile world.

The effects on our planet will be as catastrophic as they are certain. Gradually, the Earth will become hotter. In the distant future, if any of our descendants still remain, someone will experience the last perfect day on Earth. As the surface of the Sun encroaches, our oceans will boil away, the molecules in our atmosphere will be agitated off into space, and the memory of life on Earth will fade into someone's history; or perhaps no one's history if we have steadfastly remained at home.

Long after life has disappeared, the Sun will fill the horizon; it may extend beyond Earth itself. This swollen stage in a star's life is known as the Red Giant phase, marked by the final release of energy and the beginning of a long, long decline. In six billion years' time, in a most beautiful display of light and colour, our sun will shed its outer layers into space to form a planetary nebula. We know this because we have seen this sequence of events unfold in the final breath of distant stars – on someone else's sun? Written across the night sky in filamentary patches of colour are the echoes of our future.

If in the far future, somewhere in the Universe, astronomers on a world not yet formed gaze through a telescope at our planetary nebula and reflect on its beauty, they may glimpse at its heart a faintly glowing ember; all that remains of a star we once thought of as magnificent. She will be smaller than the size of Earth, less than a millionth of her current volume and a fraction of her brightness. Our sun will have become a white dwarf – the destiny of almost all the stars in our galaxy – a fading, dense remnant, momentarily masked by a colourful cloud.

If our planet survives, little more than a scorched and barren rock will remain, silhouetted darkly against the fading embers of a star.

Sirius, the brightest star in our sky, sits at just over eight light years away, which makes it one of our nearest neighbours. It is so bright that on occasion it can be observed during bright twilight, partly because of its proximity but also because it is twice as big as our sun and twenty-five times as bright. It is therefore not surprising that observations of Sirius have been recorded in the oldest of astronomical records.

For thousands of years we looked up at this beacon and assumed it was a single star, but in 1862 American astronomer Alvan Graham Clark observed a sister star hidden in the glare of Sirius's light. It took so long to notice Sirius's companion because, as the photograph taken by the Hubble Space Telescope (bottom right) reveals, it is so much dimmer than its sibling. Shining faintly in the lower left-hand corner, the small dot of light is an image of the white dwarf star Sirius B. This is one of the larger white dwarf stars discovered by astronomers, with a mass similar to our sun that is packed into a sphere the size of Earth. With no fuel left to burn, white dwarfs like Sirius B glow faintly with the residual heat of their extinguished furnaces. Like most white dwarfs, Sirius B is made primarily of oxygen and carbon (the remnants of helium fusion) packed tightly with a density a million times that of a younger, living star. This is the future of our star; a vision of the Sun's death. Slowly cooling in the freezing temperatures of deep space, it is estimated that our sun will reach this phase in around 6 billion years' time. From Earth, if indeed there is an Earth at that time, our sun will shine no brighter than a full moon on a clear night.

Death must come to all stars. One day every light in the night sky will fade and the cosmos will be plunged into eternal night. This is the most profound consequence of the arrow of time; this structured Universe that we inhabit alongside all its wonders – the stars, the planets and the galaxies – cannot last forever. As we move through the age of stars, through the aeons ahead, countless billions of stars will live and die. Eventually, though, there will be only one type of star that will remain to illuminate the Universe in its old age

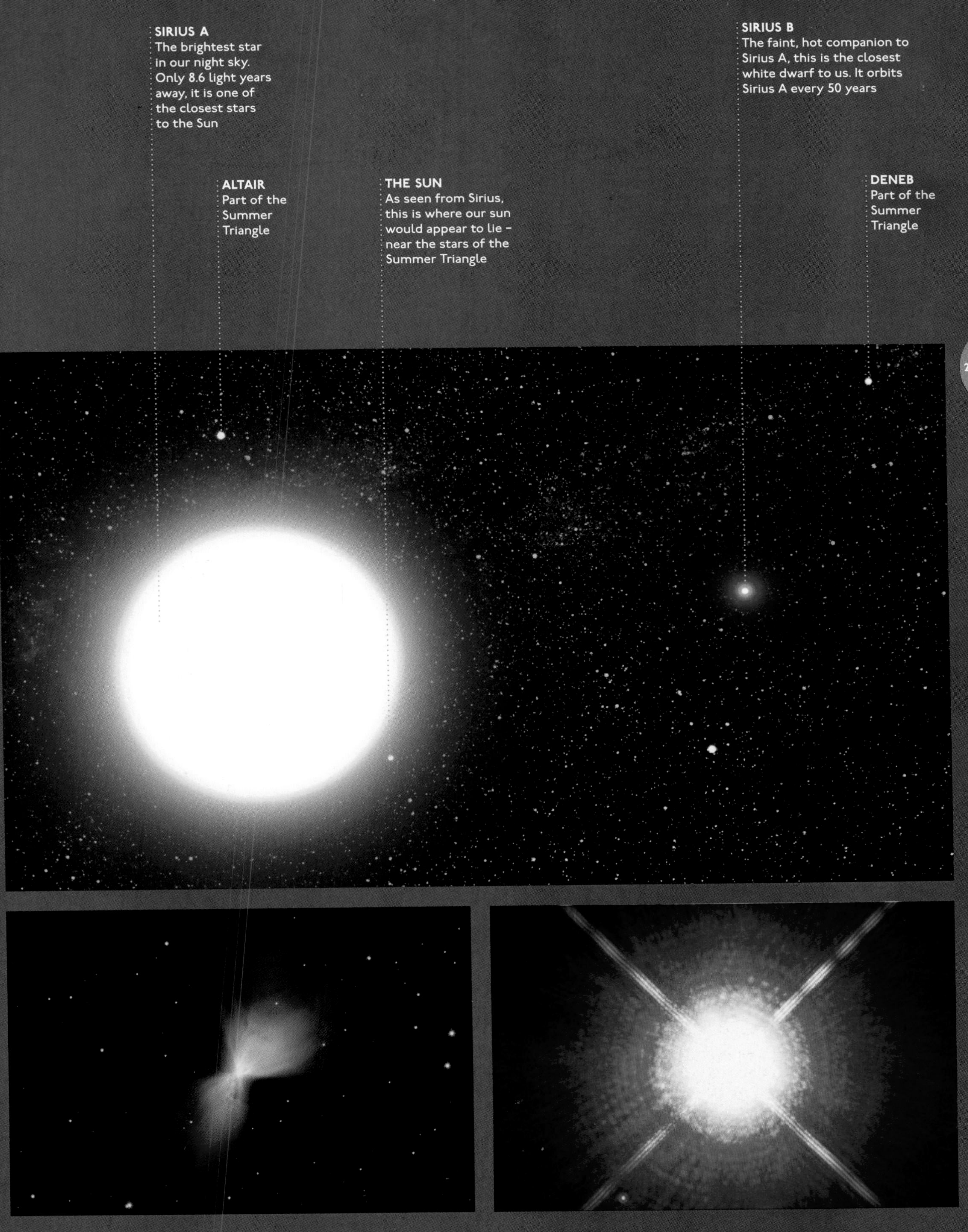
SIRIUS A
The brightest star in our night sky. Only 8.6 light years away, it is one of the closest stars to the Sun
SIRIUS B
The faint, hot companion to Sirius A, this is the closest white dwarf to us. It orbits Sirius A every 50 years
ALTAIR
Part of the Summer Triangle
THE SUN
As seen from Sirius, this is where our sun would appear to lie – near the stars of the Summer Triangle
DENEB
Part of the Summer Triangle

THE DEATH OF THE SUN

Although relatively young now, the Sun, like every other star in the Universe, must one day die. In around five billion years' time, the Sun's stores of hydrogen will run dry and the star will begin its long, dramatic swansong. During this lengthy goodbye, the last dying bursts of extra heat will extend towards us, passing Mercury and Venus on the way and leaving a trail of destruction in its wake. Long after life has disappeared on Earth, the Sun will continue to fill the horizon as it swells in the Red Giant phase until, in about six billion years' time, our Sun will shed its outer layers of gas and dust into space, exposing its core which will fade into a white dwarf, living on in the heavens as a shadow of its former self.

MERCURY

VENUS

EARTH

STAGE 1
MAIN SEQUENCE STAR

STAGE 2
RED GIANT

The Sun becomes a red giant as its stores of hydrogen run dry. Its core collapses and the extra heat this generates will cause its outer layers to expand

Outer layers of the Sun expand, cool and become redder. Its radius will increase around 250 times, taking it past Mercury, Venus and towards Earth

Earth will become increasingly hot until oceans boil away, the atmosphere escapes into space and it is left as a charred, lifeless cinder

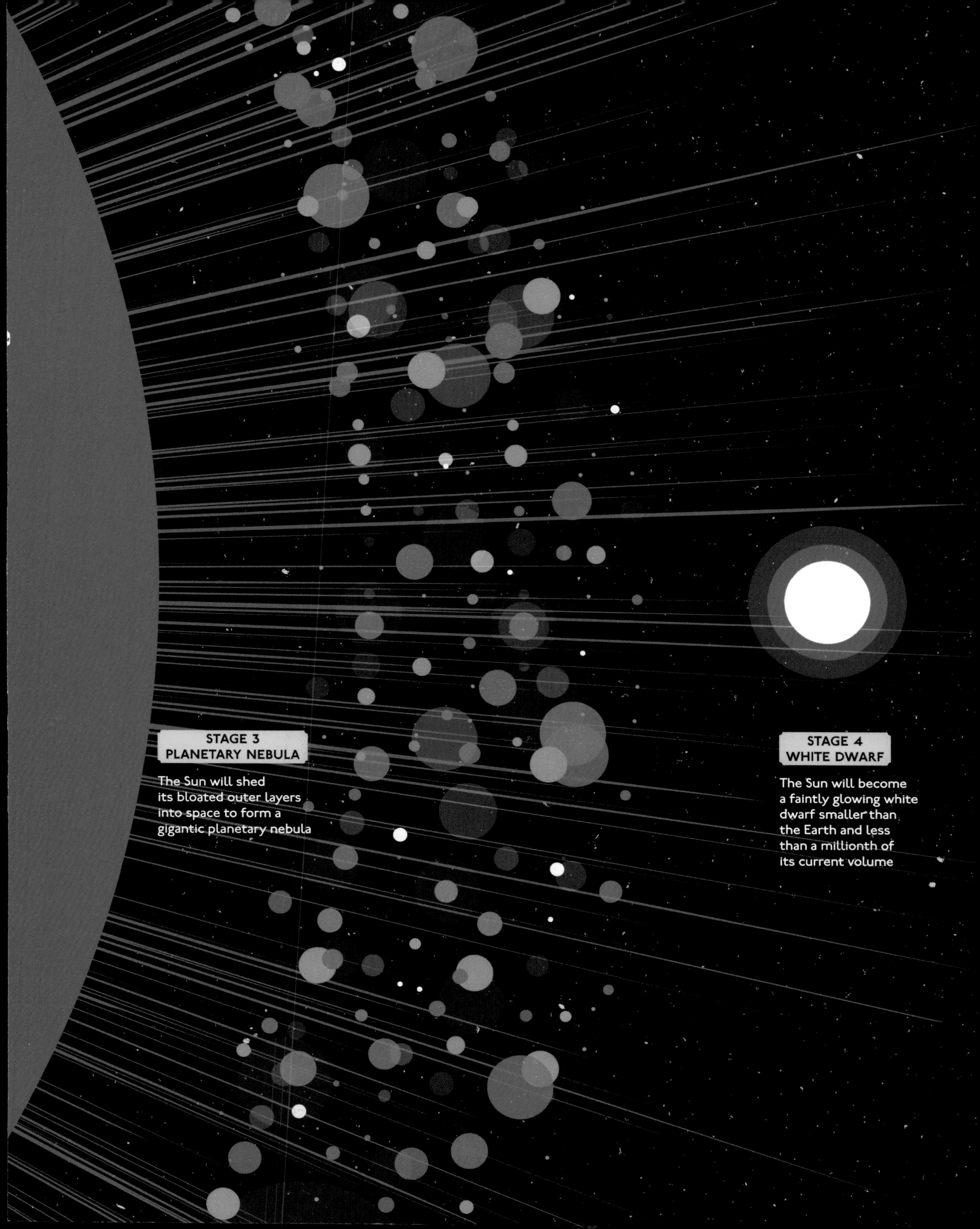

STAGE 3
PLANETARY NEBULA

The Sun will shed its bloated outer layers into space to form a gigantic planetary nebula

STAGE 4
WHITE DWARF

The Sun will become a faintly glowing white dwarf smaller than the Earth and less than a millionth of its current volume

THE LAST STARS

The nearest star to our solar system is Proxima Centauri. Although only a mere 4.2 light years away, Proxima Centauri is not visible to the naked eye from Earth and doesn't even stand out against more distant stars in many of the photographs that have been taken of it. The reason for this is that Proxima Centauri is small, very small when compared to our sun – having just 12 per cent of the Sun's mass – so to our eyes this star would appear to shine 18,000 times less brightly than our sun.

Proxima Centauri is a red dwarf star – the most common type of star in our universe. Red dwarfs are diminutive and cold, with surface temperatures in the region of 4,000K, but they do have one advantage over their more luminous and magnificent stellar brethren: because they're so small, they burn their nuclear fuel extremely slowly, and consequently they have life spans of trillions of years. This means that stars like Proxima Centauri will be the last living stars in the Universe.

If we do in fact survive into the far future of the Universe, it is possible to imagine our distant descendants building their civilisations around red dwarfs in order to capture the energy of those last fading embers of stars. Just as our ancestors crowded around campfires for warmth on cold winter nights, so some time long in the future humans may take their warmth from a red dwarf as the last available energy in the Universe.

The rate of the fusion reactions in the cores of these red dwarfs that is needed to provide the thermal pressure to resist the inward pull of their weak gravity is very low, which enables them to live longer. Even so, these are still active stars, and their surfaces are whipped up into turmoil by the turbulent convective currents that constantly churn their interiors. Amongst all this activity, explosive solar flares occur almost continually, blasting bursts of light and X-rays out into space.

Ultimately, though, the frugality of these stars is no defence against the arrow of time. Four trillion years from now, at 300 times the current age of the Universe, Proxima Centauri's fuel reserves will finally run out and the star will slowly collapse into a white dwarf. After trillions of years of stellar life and death, only white dwarfs and black holes will remain in the Universe, and then, in around 100 trillion years' time, this age of the stars will draw to a close and the cosmos will enter its next phase: The Degenerate Era. And yet, even after 100 trillion years of light, the vast majority of the Universe's history still lies ahead. Bleak, lifeless and desolate, our universe will go on, as it enters the dark ●

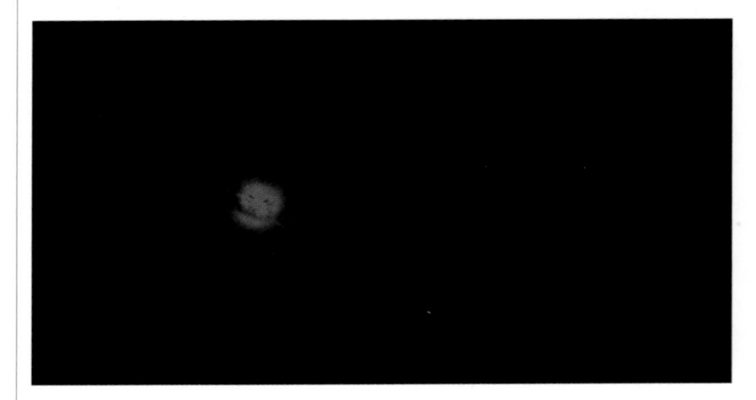

ABOVE: These computer-generated images reveal how Proxima Centauri will meet its end. Over the next four trillion years, this red dwarf will gradually collapse into a much dimmer white dwarf.

RIGHT: A white dwarf is visible amongst brighter, living stars in this enhanced image, taken by NASA's Galaxy Evolution Explorer, of Z Camelopardalis, a binary star system.

After trillions of years of stellar life and death, only white dwarfs and black holes will remain in the Universe, and then, in around 100 trillion years' time, this age of the stars will draw to a close and the cosmos will enter its next phase: The Degenerate Era.

THE BEGINNING OF THE END

On the northern coast of Namibia, where the cold waters of the South Atlantic meet the Namib Desert, lies one of the most inhospitable places on Earth. The Skeleton Coast has been feared for as long as sailors have travelled near its shores; seventeenth-century Portuguese mariners used to call this place 'the gates to hell', and the native Namib Bushmen named it 'the land God made anger in'. Today, you can just about make it to the coast in a sturdy 4x4, or effortlessly cruise in from the port city of Walvis Bay in a helicopter. But even so, when you stand on the sands beside the South Atlantic, the gods still have anger left. Each morning a dense ocean fog rolls along the coastline, fed by the upwelling of the cold Benguela current. Coupled with the constantly shifting shape of the sandbanks in the intense Atlantic winds, this toxic navigational conspiracy has meant that over the years thousands of ships have been wrecked along the Skeleton Coast. The decaying carcasses of the rusting ships and the bleached bones of marine life swept ashore by the currents all add to the coast's gothic feel. The name Skeleton Coast also reflects the large number of human lives lost here over the centuries; even if you made it ashore after a shipwreck, the onshore currents are so strong that there is no way of rowing back out to sea, and the only route to safety is through hundreds of miles of inhospitable desert. This genuinely was a place of no return: if you were shipwrecked here, this was the end of your universe.

One of the ships to end her days here was the Eduard Bohlen, a 91-metre (300-foot), 2,272-tonne steamship that ran aground here on 5 September 1909 on a journey from Germany to West Africa. A century's shifting sands have carried her hundreds of metres inland and the Atlantic winds have attacked her carcass, leaving her rusting and skeletal. When we arrive she is guarded by a phalanx of jackals who are less wary of us than I expected. She forms an abstract

Just as the ship's iron will eventually rust and be carried away by the desert winds, so we think the last matter in the Universe will eventually be carried off into the void.

BELOW: The Skeleton Coast: one of the most inhospitable places on Earth, where humans have perished for centuries, and where only jackals and the strongest life forms remain.

backdrop to our story; the symbolism is immediate, brutal even, and for me surprisingly powerful. These wrecks, complex structures dismantled by the passage of time, are like our last stars.

In the far future of the cosmos, the last remaining beacons of light will no more be permitted to evade the second law of thermodynamics than the Eduard Bohlen. Even the white dwarfs must fade as the laws of physics methodically dismantle the Universe. Slowly, as the glowing embers of the last stars lose their warmth to space, they will cease to emit visible light. After trillions of years, the final beacons burning in the cosmic sky will turn cold and dark – their remnants are known as black dwarfs.

Black dwarfs are dark, dense, decaying balls of degenerate matter. Nothing more than the ashes of stars, they take so long to form that after almost 14 billion years, the Universe is currently too young to contain any at all. Yet despite never seeing one, our understanding of fundamental physics allows us to make concrete predictions about how they will end their days. Just as the iron that makes up the ships of the Skeleton

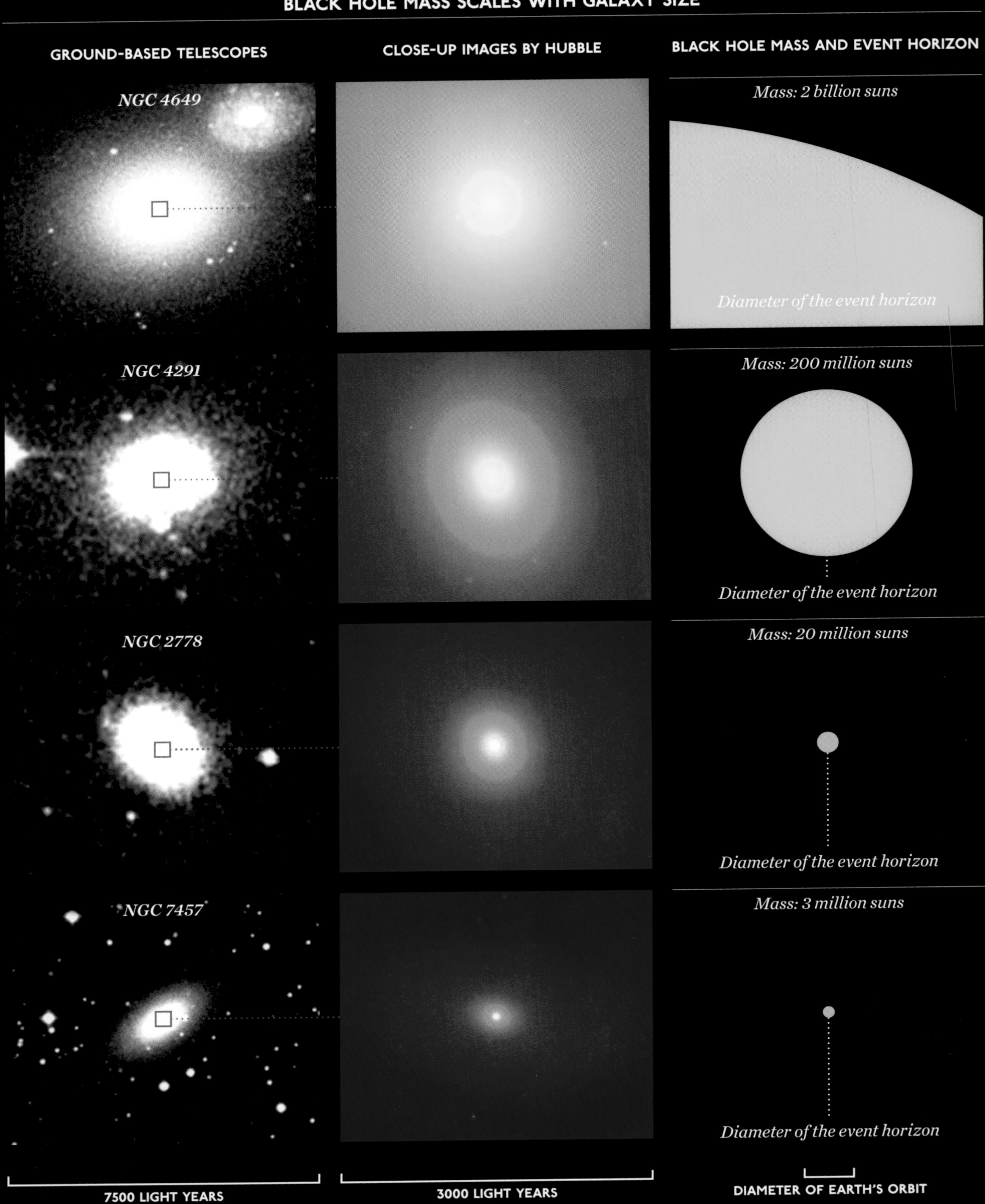
BLACK HOLE MASS SCALES WITH GALAXY SIZE
GROUND-BASED TELESCOPES
CLOSE-UP IMAGES BY HUBBLE
BLACK HOLE MASS AND EVENT HORIZON
NGC 4649
Mass: 2 billion suns
Diameter of the event horizon
NGC 4291
Mass: 200 million suns
Diameter of the event horizon
NGC 2778
Mass: 20 million suns
Diameter of the event horizon
NGC 7457
Mass: 3 million suns
Diameter of the event horizon
7500 LIGHT YEARS
3000 LIGHT YEARS
DIAMETER OF EARTH'S ORBIT

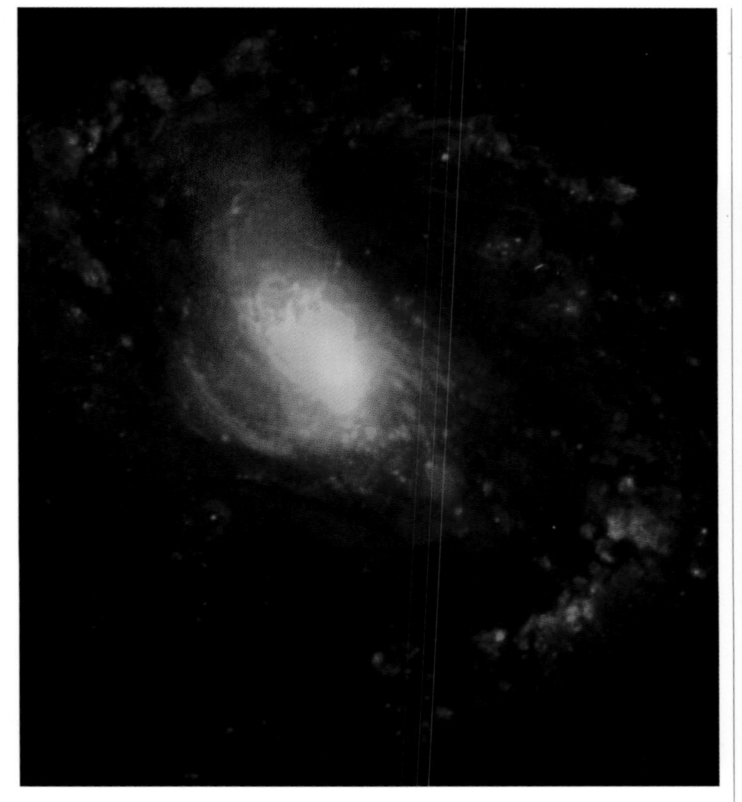

LEFT: This composite X-ray image from the Chandra X-ray Observatory shows gas blowing away from a central supermassive black hole in the active galaxy NGC 1068.

BELOW: In trillions of years, our universe will be littered with black dwarfs. From the ashes of stars, dark, dense and decaying balls of degenerate matter will form.

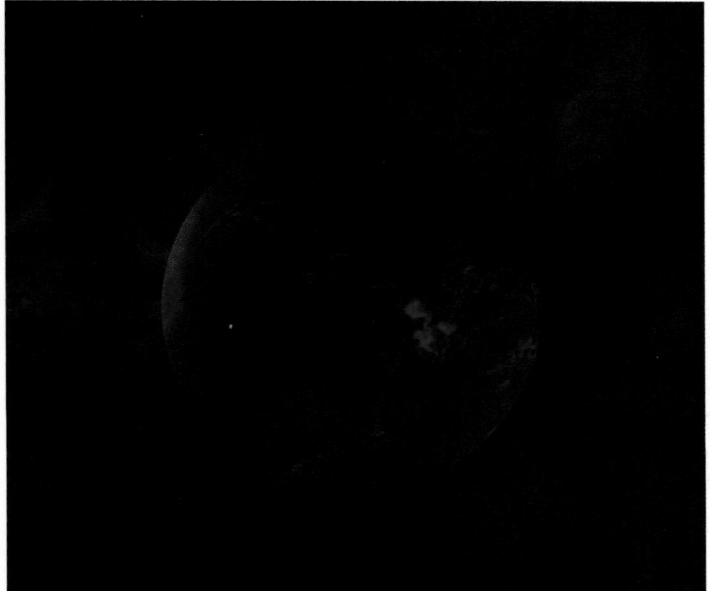

Once the last remnants of the last stars have decayed away to nothing . . . the story of our universe will finally come to an end.

Coast will eventually be carried away by the desert winds, so it is thought that the matter inside black dwarfs, the last matter in the Universe, will eventually evaporate away and be carried off into the void as radiation, leaving nothing behind. The processes by which matter might, given enough time, decay, are not understood. Physicists need a more advanced theory of the forces of nature, known as a Grand Unified Theory, to speak with certainty about the behaviour of protons, neutrons and electrons over trillion-year timescales. There are reasons to expect that such a theory may exist, and that a mechanism for even the most stable sub-atomic particles to decay into radiation might be present in nature. For this reason, experiments to measure the lifetime of protons are ongoing in laboratories around the world, but as yet nobody has observed proton decay, and we are therefore now in the realm of speculation. But here is one possible, and given our understanding of physics today, probable, story of how our universe will end.

With the black dwarfs gone, there will not be a single atom of matter left in the Universe. All that will remain of our once-rich cosmos will be particles of light and black holes. After an unimaginable expanse of time, it is thought that even the black holes will evaporate away, and the Universe will consist of a sea of light; photons all tending to the same temperature as the expansion of the Universe cools them towards absolute zero. When I say unimaginable period of time, I really mean it: ten thousand trillion trillion trillion trillion trillion trillion trillion trillion years. In scientific notation, that's 10^{100} years. That is a very big number indeed; if I were to start counting with a single atom representing one year, there wouldn't be enough atoms in all the stars and planets in all the galaxies in the entire observable universe to get anywhere near that number.

Once the last remnants of the last stars have decayed away to nothing and everything reaches the same temperature, the story of our universe will finally come to an end. For the first time in its life the Universe will be permanent and unchanging. Entropy finally stops increasing because the cosmos cannot get any more disorganised. Nothing happens, and it keeps not happening forever.

This is known as the heat death of the Universe, an era when the cosmos will remain vast, cold, desolate and unchanging for the rest of time. There's no way of measuring the passing of time, because nothing in the cosmos changes. Nothing changes because there are no temperature differences, and therefore no way of moving energy around to make anything happen. The arrow of time has simply ceased to exist. This is an inescapable fact, written into the fundamental laws of physics. The cosmos will die; every single one of the hundreds of billions of stars in the hundreds of billions of galaxies in the Universe will expire, and with them any possibility of life in the Universe will be extinguished ◉

A VERY PRECIOUS TIME

The fact that the Sun will die, incinerating Earth and obliterating all life on our planet, and that eventually the rest of the stars in the Universe will follow suit to leave a vast, formless cosmos with no possibility of supporting any life or retaining any record of the living things that brought meaning to its past, might sound a bit depressing to you. You might legitimately ask questions about the way our universe is put together. Surely you could build a universe in a different way? Surely you build a universe such that it didn't have to descend from order into chaos? Well, the answer is 'no', you couldn't, if you wanted life to exist in it.

The arrow of time, the sequence of changes that will slowly but inexorably lead the Universe to its death, is the very thing that created the conditions for life in the first place. It took time for the Universe to cool sufficiently after the Big Bang and for matter to form; it took time for gravity to clump the matter together to form galaxies, stars and planets, and it took time for the matter on our planet to form the complex patterns that we call life. Each of these steps took place in perfect accord with the Second Law of Thermodynamics; each is a step on the long road from order to disorder.

The arrow of time has created a bright window in the Universe's adolescence during which life is possible, but it's a window that won't stay open for long. As a fraction of the lifespan of the Universe, as measured from its beginning to the evaporation of the last black hole, life as we know it is only possible for one-thousandth of a billion billion billionth, billion billion billionth, billion billion billionth of a per cent.

And that's why, for me, the most astonishing wonder of the Universe isn't a star or a planet or a galaxy; it isn't a thing at all – it's a moment in time. And that time is now.

Around 3.8 billion years ago life first emerged on Earth; two hundred thousand years ago the first humans walked the plains of Africa; two and a half thousand years ago humans believed the Sun was a god and measured its orbit with stone towers built on the top of a hill. Today, our curiosity manifests itself not as sun gods but as science, and we have observatories – almost infinitely more sophisticated than the Thirteen Towers – that can gaze deep into the Universe. We have witnessed its past and now understand a significant amount about its present. Even more remarkably, using the twin disciplines of theoretical physics and mathematics, we can

ABOVE: This colour image of the Earth, named the 'Pale Blue Dot', is a part of the first-ever portrait of the Solar System taken by NASA's Voyager 1. The spacecraft took 60 frames which could be used to create a mosaic image of the Solar System from a distance of over four billion miles from Earth.

ABOVE: This seemingly insignificant image of a pale blue dot is in fact one of the most important and beautiful images ever taken, revealing our planet at a distance of over six billion kilometres away.

RIGHT: Our time on Earth is precious and fleeting. The most important use of this time that we can make is to ask questions about our wonderful universe, so that perhaps one day one of our descendants will truly understand the natural laws that govern our cosmos.

calculate what the Universe will look like in the distant future and make concrete predictions about its end.

I believe it is only by looking out to the heavens, by continuing our exploration of the cosmos and the rules that govern it, and by allowing our curiosity free reign to wander the limitless natural world, that we can understand ourselves and our true significance within this Universe of wonders.

In 1977, a space probe called Voyager 1 was launched on a 'grand tour' of the Solar System. It visited the great gas giant planets Jupiter and Saturn and made wonderful discoveries before heading off into interstellar space. Thirteen years later, after its mission was almost over, Voyager turned its cameras around and took one last picture of its home. This picture (left) is known as the Pale Blue Dot. The beautiful thing, perhaps the most beautiful thing ever photographed, is the single pixel of light at its centre; because that pixel, that point, is our planet, Earth. At a distance of over six billion kilometres (3.7 billion miles) away, this is the most distant picture of our planet that has ever been taken.

The powerful and moving thing about this tiny, tiny point of light is that every living thing that we know of that has ever existed in the history of the Universe has lived out its life on that pixel, on a pale blue dot hanging against the blackness of space.

As the great astronomer Carl Sagan wrote:

'It has been said that astronomy is a humbling and character-building experience. There is perhaps no better demonstration of the folly of human conceits than this distant image of our tiny world. To me, it underscores our responsibility to deal more kindly with one another, and to preserve and cherish the pale blue dot, the only home we've ever known.'

Just as we, and all life on Earth, stand on this tiny speck adrift in infinite space, so life in the Universe will only exist for a fleeting, dazzling instant in infinite time, because life, just like the stars, is a temporary structure on the long road from order to disorder.

But that doesn't make us insignificant, because life is the means by which the Universe can understand itself, if only for an instant. This is what we've done in our brief moments on Earth: we have sent space probes to the edge of our solar system and beyond; we have built telescopes that can glimpse the oldest and most distant stars, and we have discovered and understood at least some of the natural laws that govern the cosmos. This, ultimately, is why I believe we are important. Our true significance lies in our continuing desire to understand and explore this beautiful Universe – our magnificent, beautiful, fleeting home ◉

'Somewhere, something incredible is waiting to be known'

— Carl Sagan, 1934–1996

INDEX

Entries in *italics* indicate photographs and images

D

F

I

L

N

S

T

U

ACKNOWLEDGEMENTS

In writing this book we'd like to thank all of those who were involved in the BBC television production of *Wonders of the Universe*. We'd especially like to thank Jonathan Renouf and James van der Pool for their commitment and dedication to the series and Stephen Cooter, Michael Lachmann and Chris Holt for transforming such complex content into beautiful television.

We'd like to thank, Rebecca Edwards, Diana Ellis-Hill, Laura Mulholland, Ben Wilson, Kevin White, George McMillan, Chris Openshaw, Darren Jonusas, Peter Norrey, Simon Sykes, Suzie Brand, Louise Salkow, Laura Davey, Paul Appleton, Sheridan Tongue, Julie Wilkinson, Laetitia Ducom , Lydia Delmonte, Daisy Newman, Jane Rundle, Nicola Kingham and the team at BDH and Unit post production.

We'd like to thank Sue Ryder, Professor Jeff Forshaw, Myles Archibald and all the team at Harper Collins for their help and guidance. Thanks also to Peter Hubbard and Marta Schooler for bringing the book to the USA.

We'd like to thank Kevin White for his outstanding photography on location.

Brian would like to thank The University of Manchester and The Royal Society for allowing him the time to make *Wonders*.

Andrew would like to thank Anna for her endless support in the writing of this book.

PICTURE CREDITS

The authors and publisher would like to thank Burrell Durrant Hifle for all the CGI images and for their help with the project as well as Jon Murray at UNIT for his help.

All pictures are copyright of the BBC except:

22–23, 62–63, 88–89, 108–109, 110–111, 128–129, 196–197, 206–207, 222–223, 232–233 Nathalie Lees © HarperCollins; 24, 25, 26, 27, 29 (bottom), 30, 37, 41, 49, 51, 52, 53, 54, 61, 67, 69 (bottom), 70, 74–75, 76, 83, 84, 85, 96, 97, 120, 124, 125, 132, 133 (bottom), 138, 141, 142, 143, 149 (top), 155 (bottom), 161, 167, 169, 176, 180, 181, 184, 191, 198, 227, 231, 235, 238, 239 (left), 240 (left) NASA; 34 GIPHOTOSTOCK / SCIENCE PHOTO LIBRARY; 39 © Scott Smith/Corbis; 45 © JASON REED/Reuters/CORBIS; 48 (left) PASCAL GOETGHELUCK / SCIENCE PHOTO LIBRARY; 48 (right) AlltheSky.com; 79 DAVID PARKER / SCIENCE PHOTO LIBRARY; 87 ECKHARD SLAWIK / SCIENCE PHOTO LIBRARY; 93 (bottom) DIRK WIERSMA / SCIENCE PHOTO LIBRARY; 94–95 © 2010 Theodore Gray; 103 © Charles O'Rear/ CORBIS; 107 CERN / SCIENCE PHOTO LIBRARY; 115 US DEPARTMENT OF ENERGY / SCIENCE PHOTO LIBRARY; 119 © Tony Hallas/Science Faction/Corbis; 146– 147 © NASA/Corbis; 148 NASA / SCIENCE PHOTO LIBRARY; 149 (bottom) Caltech; 151 ERICH SCHREMPP / SCIENCE PHOTO LIBRARY; 156–157 ESA / DLR / FU BERLIN (G.NEUKUM) / SCIENCE PHOTO LIBRARY; 164–165 © Tony Hallas/Science Faction/Corbis; 170–171 John Dubinski; 173 B. MCNAMARA (UNIVERSITY OF WATERLOO) / NASA / ESA/ STScI / SCIENCE PHOTO LIBRARY; 174–175 © Roger Ressmeyer/CORBIS; 183 MARTIN BOND / SCIENCE PHOTO LIBRARY; 186–187 SCIENCE PHOTO LIBRARY; 188 ROYAL ASTRONOMICAL SOCIETY / SCIENCE PHOTO LIBRARY; 194 NASA / SCIENCE PHOTO LIBRARY; 209 SCIENCE PHOTO LIBRARY; 215 SCIENCE, INDUSTRY & BUSINESS LIBRARY / NEW YORK PUBLIC LIBRARY / SCIENCE PHOTO LIBRARY; 242-243 © CORBIS.